Mathematical Physics II.
Classical Statistical Mechanics.
Lecture Notes

Matteo Petrera

Institut für Mathematik, MA 7-2
Technische Universität Berlin
Strasse des 17. Juni 136

Bibliographic information published by the Deutsche Nationalbibliothek

The Deutsche Nationalbibliothek lists this publication in the Deutsche Nationalbibliografie; detailed bibliographic data are available in the Internet at http://dnb.d-nb.de .

ISBN 978-3-8325-3719-7

Logos Verlag Berlin GmbH
Comeniushof, Gubener Str. 47,
10243 Berlin
Tel.: +49 (0)30 42 85 10 90
Fax: +49 (0)30 42 85 10 92
INTERNET: http://www.logos-verlag.de

Motivations

These Lecture Notes are based on a one-semester course taught in the Summer semesters of 2012 and 2014 at the Technical University of Berlin, to bachelor undergraduate mathematics and physics students. The exercises at the end of each chapter have been either solved in class, during Tutorial hours, or assigned as weekly homework.

These Lecture Notes are based on the references listed on the next pages. It is worthwhile to warn the reader that there are several excellent and exhaustive books and monographs on the topics that were covered in this course. A practical drawback of some of these books is that they are not really suited for a 4-hours per week, one-semester course. On the contrary, these notes contain only those topics that were actually explained in class. They are a kind of one-to-one copy of blackboard lectures. Some topics, some aspects of the theory and some proofs were left out because of time constraints.

A characteristic feature of these notes is that they present subjects in a synthetic and schematic way, thus following exactly the same pedagogical strategy used in class. Notions, concepts, statements and proofs are intentionally written and organized in a way that I found well suited for a systematic and effective understanding/learning process.

The aim is to provide students with practical tools that allow them to prepare themselves for their exams and not to substitute the role of an exhaustive book. This purpose has, of course, drawbacks and benefits at the same time. As a matter of fact, many students wish to have a "product" which is readable, compact and self-contained. In other words, something that is necessary and sufficient to get a good mark with a reasonable effort. This is - at least ideally - the positive side of good Lecture Notes. On the other hand, the risk is that their understanding might not be fluid and therefore too confined. Indeed, I always encourage my students to also consult more "standard" books like the ones quoted on the next pages.

Aknowledgments

The DFG (Deutsche Forschungsgemeinschaft) collaborative Research Center TRR 109 "Discretization in Geometry and Dynamics" is acknowledged. I am grateful to Yuri Suris and to the BMS (Berlin Mathematical School) for having given me the opportunity to teach this course. I thank Andrea Tomatis for his careful proofreading.

Matteo Petrera
Institut für Mathematik, MA 7-2, Technische Universität Berlin
Strasse des 17. Juni 136
petrera@math.tu-berlin.de
August 26, 2014

Books and references used during the preparation of these Lecture Notes

Ch1 Motivations and Background

- ✓ [Bo] A. Bovier, *Lecture Notes. Gibbs Measures and Phase Transitions*, available at http://wt.iam.uni-bonn.de/bovier/home/.
- ✓ [FaMa] A. Fasano, S. Marmi, *Meccanica Analitica: una Introduzione*, Bollati Boringhieri, 2002.

Ch2 Introduction to Kinetic Theory of Gases

- ✓ [FaMa] A. Fasano, S. Marmi, *Meccanica Analitica: una Introduzione*, Bollati Boringhieri, 2002.
- ✓ [GoOl] G.A. Gottwald, M. Oliver, *Boltzmann's Dilemma: An Introduction to Statistical Mechanics via the Kac Ring*, SIAM Review, 51/3, 2009.
- ✓ [Hu] K. Huang, *Statistical Mechanics*, Wiley & Sons, 1987.
- ✓ [Th] C.J. Thompson, *Mathematical Statistical Mechanics*, MacMillan, 1972.
- ✓ [Ue] D. Ueltschi, *Introduction to Statistical Mechanics*, available at http://www.ueltschi.org/teaching/2006-MA4G3.html.

Ch3 Gibbsian Formalism for Continuous Systems at Equilibrium

- ✓ [Bo] A. Bovier, *Lecture Notes. Gibbs Measures and Phase Transitions*, available at http://wt.iam.uni-bonn.de/bovier/home/.
- ✓ [FaMa] A. Fasano, S. Marmi, *Meccanica Analitica: una Introduzione*, Bollati Boringhieri, 2002.
- ✓ [Ga] G. Gallavotti, *Statistical Mechanics: a Short Treatise*, Springer, 1999.
- ✓ [Hu] K. Huang, *Statistical Mechanics*, Wiley & Sons, 1987.
- ✓ [Kh] A.I. Khinchin, *Mathematical Foundations of Statistical Mechanics*, Dover, 1949.
- ✓ [LaLi] L.D. Landau, E.M. Lifshitz, *Statistical Physics*, Pergamon Press, 1969.
- ✓ [Mo] G. Morandi, *Statistical Mechanics: An Intermediate Course*, World Scientific, 1996.
- ✓ [Th] C.J. Thompson, *Mathematical Statistical Mechanics*, MacMillan, 1972.

Ch4 Introduction to Ising Models

- ✓ [McC] B. McCoy, *Advanced Statistical Mechanics*, Oxford University Press, 2010.
- ✓ [McCWu] B. McCoy, T.T. Wu, *The two-dimensional Ising Model*, Harvard University Press, 1973.
- ✓ [Hu] K. Huang, *Statistical Mechanics*, Wiley & Sons, 1987.
- ✓ [Ka] B. Kaufman, *Crystal Statistics II. Partition Function Evaluated by Spinor Analysis*, Phys. Rev. 76/8, 1949.
- ✓ [LaBe] D. Lavis, G. Bell, *Statistical Mechanics of Lattice Systems Volume 1: Closed-Form and Exact Solutions*, Springer, 1999.
- ✓ [Th] C.J. Thompson, *Mathematical Statistical Mechanics*, MacMillan, 1972.

James Clerk Maxwell (1831-1879) and Ludwig Eduard Boltzmann (1844-1906)

Josiah Willard Gibbs (1839-1903)

From "The value of science" (1905) by Jules Henri Poincaré (1854-1912): *"A drop of wine falls into a glass of water; whatever may be the law of the internal motion of the liquid, we shall soon see it colored of a uniform rosy tint, and however much from this moment one may shake it afterwards, the wine and the water do not seem capable of again separating. Here we have the type of the irreversible physical phenomenon: to hide a grain of barley in a heap of wheat, this is easy; afterwards to find it again and get it out, this is practically impossible. All this Maxwell and Boltzmann have explained; but the one who has seen it most clearly, in a book too little read because it is a little difficult to read, is Gibbs, in his* Elementary Principles of Statistical Mechanics.*"*

Contents

1

Motivations and Background

1.1 Introduction

▶ On the basis of Newtonian mechanics the 18th and 19th century have seen with the development of analytical mechanics a powerful and effective tool for the analysis and prediction of natural phenomena.

- Completely integrable Hamiltonian systems are the mechanical models for the study of systems with a regular and completely predictable behavior. The main idea in all studies of the 19th century has been to reduce the study of mechanical systems to the study of integrable systems, both exactly (canonical transformations theory, Hamilton-Jacobi theory) and approximately (Hamiltonian perturbation theory). Poincaré was the first to prove in a rigorous way that there exist mechanical systems which may exhibit a behavior that is totally different from the behavior of integrable systems, exhibiting disorderly and chaotic orbits. The appropriate language for the study of these systems connects dynamical systems theory to probability theory. This is the point of view underlying *ergodic theory*.

- At the same time new areas of physics became the target of new research. One of the most important of these new fields was the theory of heat, or *thermodynamics*. Thermodynamics was probably the first research area which needed the introduction of some innovative concepts and quantities whose set-up was not only Newtonian mechanics. One of the main principles of Newtonian mechanics is the *conservation of energy*. For real systems such a principle could not hold entirely, due to the ubiquitous dissipation of energy. As a matter of fact all machines need some source of energy, for instance heat. A central objective of thermodynamics was to understand how the two types of energy, mechanical and thermal, could be converted into each other.

- Thermodynamics was originally a pragmatic theory: the new concepts related to the phenomenon of heat, *temperature* and *entropy*, were coupled with the mechanical concepts of energy and force. Towards the end of the 19th century Boltzmann proposed a mechanical interpretation of thermodynamic effects in terms of atomistic theory. This *kinetic theory of gases* turned into what we now know as (classical) *statistical mechanics* through the work of Gibbs in the early 20th century.

▶ Statistical mechanics has originated from the desire to obtain a mathematical understanding of a class of of physical systems of the following nature:

1

- The system is an assembly of identical subsystems subject to some macroscopic constraints and boundary conditions.

- The number of subsystems is large.

- The interactions between subsystems produce a phenomenological and macroscopic *thermodynamic behavior* of the system:

    (a) A *thermodynamic state* is completely and uniquely defined by values of a suitable set of parameters known as *thermodynamic variables* (pressure, temperature, volume, etc.). Such variables allow us to define some *thermodynamic functions* (entropy, energy, free energy, etc.).

    (b) *Equilibrium thermodynamic states* are those states corresponding to *thermodynamic equilibrium*, which are characterized by zero flows of all quantities, both internal and between the system and surroundings. An equilibrium state of a system consists of one or more macroscopic homogeneous regions, calles *thermodynamic phases*. It is supposed that at equilibrium thermodynamic functions depend smoothly on thermodynamic variables. If some singularities of thermodynamic functions occur then they correspond to severe changes in the phase structure of the system (*phase transitions*).

    (c) For non-equilibrium thermodynamics, a suitable set of state variables includes some macroscopic quantities which exhibit departure from thermodynamic equilibrium. Non-equilibrium states indicate the existence of some non-vanishing flow within the system or between the system and surroundings.

    (d) The state of an isolated system tends to an equilibrium state as time goes to infinity. Once the boundary conditions have been specified, there is usually one and only one final equilibrium state towards which the system evolves. Once it has been attained within some characteristic relaxation time, the system will stay there forever (unless we change the boundary conditions, of course), never (in statistical sense) returning to its initial state and accomplishing only small fluctuations around the equilibrium state.

The prototypical example of a statistical mechanical system is a gas of particles subject to some macroscopic constraints, as for instance a volume containing a large number of particles.

▶ The purpose of this Chapter is twofold:

- To give some concrete physical motivations for the study of statistical mechanics.

- To present - in a very elementary fashion - some notions and tools, both physical and mathematical, which will be used in the rest of the course.

1.2 A few words about thermodynamics

▶ As a matter of fact, there are some intrinsic features of real objects, beyond position and velocity, that may interfere with their mechanical properties. An example is given by the *temperature*. Thermodynamics introduces a description of such internal variables and devises a theory allowing to control the associated flows of energy.

▶ The classical setting of thermodynamics is a *gas*, namely a huge collection of small particles (molecules) enclosed in a container of a given but possibly variable *volume* V.

- This container provides the means to couple the system to an external mechanical system: if one can make the gas change V, the resulting motion can be used to power a machine. Conversely, we may change V and thus change the properties of the gas inside.

- The (positive) parameter which describes the state of the gas that reacts to the change of V is the *pressure*, P. The definition of the pressure is given in terms of the amount of mechanical energy needed to change the volume:

$$P := -\frac{dE_{mech}}{dV}. \tag{1.1}$$

- Formula (1.1) must depend on further parameters. An obvious one is the total amount of gas in the container, i.e., the number of particles N. However it is natural to assume that, if $V = Nv$, where v is a given specific volume, then P should not depend explicitly on N. For this reason one says that the pressure is an *intensive variable*. By contrast, V is an *extensive variable*. It follows that E_{mech} is extensive.

- Extensive variables are those variables which scale with the volume of the system. Consider two subsystems with energies E_1 and E_2 and put them together. Let E be the energy of the resulting system. A reasonable definition of the total energy being extensive claims that the ratio $E/(E_1 + E_2)$ should tend to one as volumes of the subsystems tend to infinity (if surface effects become negligible in that limit). This statement contains an intuitive definition of the notion of *thermodynamic limit*.

- The number of molecules N is not necessarily constant and its change may involve a change of energy, which is of chemical origin. The parameter which governs this energy change is the *chemical potential*

$$\mu := \frac{dE_{chem}}{dN}.$$

- In order to have total energy conservation, we must take into account a further internal variable property of the gas. This extensive and positive definite quantity is called *entropy*, denoted by S. The (absolute) *temperature*, T, is the positive intensive variable that relates its change to the change of energy. Traditionally, this thermal energy is called *heat* and denoted by Q, so that we define

$$T := \frac{dQ}{dS}. \tag{1.2}$$

- The principle of conservation of energy then states that any change of the parameters of the system is such that the *first law of thermodynamics* is satisfied:

$$dE_{\text{mech}} + dE_{\text{chem}} + dQ = dE, \tag{1.3}$$

  namely

$$-P\,dV + \mu\,dN + T\,dS = dE. \tag{1.4}$$

  The total energy of the system, E, is called *internal energy*.

- One postulates that the equilibrium thermodynamic state is described by giving the value of the three extensive variables V, N, S. Therefore one assumes that the thermodynamic phase space is a three-dimensional manifold. In particular, one defines the total internal energy

$$E(V, N, S) := E_{\text{mech}} + E_{\text{chem}} + Q,$$

  and correspondingly the following intensive variables at equilibrium:

$$P := -\frac{\partial E}{\partial V}, \qquad \mu := \frac{\partial E}{\partial N}, \qquad T := \frac{\partial E}{\partial S}. \tag{1.5}$$

- Equations (1.5) are called *equations of state at equilibrium*. Suppose we fix the intensive variables P, μ, T to certain values, and set the extensive variables V, N, S to some initial values. Then the time evolution of the system will drive these parameters to equilibrium, i.e., to values for which equations (1.5) hold. Such processes are *irreversible*. In contrast, a *reversible* process varies intensive and extensive parameters in such a way that it passes along equilibrium states and the equations of state (1.5) hold both in the initial and in the final state of the process.

- The notion of irreversibility of macroscopic processes is encoded in the *second law of thermodynamics*, according to which the entropy S "never" decreases in any "natural" process, i.e., in any process taking place when the system is in isolation. Here "isolation" specifies the boundary conditions and equilibrium under such conditions is characterized by the entropy of the system being maximized.

- A characteristic feature of thermodynamics is the possibility to re-express the equations of state in terms of different sets of variables, e.g., to express T, N, S as a function of P, V, T, etc. To ensure that this is possible, one always assumes that E is a (strictly) convex function. Then, the desired change of variables can be achieved with the help of Legendre transformations.

1.3 A few words about ergodic dynamical systems

▶ We will see that the so called *ergodic hyphotesis* plays a central role in statistical mechanics of continuous systems. In order to understand the meaning of ergodicity of a dynamical system we need some facts from measure theory and probability theory.

1.3.1 *Measure spaces and measurable functions*

▶ Let X be a non-empty set.

- A non-empty family $\mathscr{A}$ of subsets of X is a *σ-algebra* on X if:

  (a) $A \in \mathscr{A}$ implies that $A^c := X \setminus A \in \mathscr{A}$.
  (b) For any sequence $\{A_i\}_{i \in \mathbb{N}}$, $A_i \in \mathscr{A}$, there holds $\bigcup_{i \in \mathbb{N}} A_i \in \mathscr{A}$.

  It is immediate to verify that any σ-algebra is also an algebra. In particular if $A, B \in \mathscr{A}$ then $A \cup B \in \mathscr{A}$.

- The following facts are true:

  1. Let $\mathscr{A}$ be a σ-algebra on X. Condition (b) is equivalent to

     $$\bigcap_{i \in \mathbb{N}} A_i = \left(\bigcup_{i \in \mathbb{N}} A_i^c \right)^c \in \mathscr{A}.$$

     In particular $A \cap B \in \mathscr{A}$ for all $A, B \in \mathscr{A}$.
  2. Let $\mathscr{A}$ be a σ-algebra on X. Then $\varnothing \in \mathscr{A}, X \in \mathscr{A}$. Indeed $A \in \mathscr{A}$ implies $X = A \cup A^c \in \mathscr{A}, \varnothing = A \cap A^c \in \mathscr{A}$.
  3. Let $\mathscr{A}$ be a σ-algebra on X. If $A, B \in \mathscr{A}$ then $A \setminus B := A \cap B^c \in \mathscr{A}$.
  4. The intersection of σ-algebras on X is a σ-algebra on X.
  5. The family of all subsets of X, called the *power set* of X and denoted by $\mathscr{P}(X)$, is a a σ-algebra.

- Property 4 allows one to generate the smallest σ-algebra on X containing a prescribed family $\mathscr{F}$ of subsets of X. Given a family $\mathscr{F}$ of subsets of X the σ-algebra on X *generated by* $\mathscr{F}$ is the intersection of all σ-algebras $\mathscr{A}$ such that $\mathscr{A} \supset \mathscr{F}$. This definition is meaningful because there exists at least one σ-algebra $\mathscr{A}$ such that $\mathscr{A} \supset \mathscr{F}$, the σ-algebra of all subsets of X.

- Let X be a topological space. The *Borel σ-algebra of X*, denoted by $\mathscr{B}(X)$, is the σ-algebra generated by the open subsets of X. The elements of $\mathscr{B}(X)$ are called *Borel sets of X*.

▶ A measure on X is a function which assigns a non-negative real number to subsets of $\mathscr{A}$. This can be thought of as making the notion of "size" or "volume" for sets into a precise concept. We want the size of the union of disjoint sets to be the sum of their individual sizes, even for an infinite sequence of disjoint sets. We now give a formal definition of measure and measurable space.

Definition 1.1

Let $\mathscr{A}$ be a σ-algebra on X. A **measure** is a function $\mu : \mathscr{A} \to [0, +\infty]$, such that:

1. $\mu(\varnothing) = 0$.

2. For any sequence $\{A_i\}_{i \in \mathbb{N}}$ of disjoint elements of $\mathscr{A}$ there holds:

$$\mu \left(\bigcup_{i \in \mathbb{N}} A_i \right) = \sum_{i \in \mathbb{N}} \mu(A_i).$$

The pair $(X, \mathscr{A})$ is called **measurable space** and the triple $(X, \mathscr{A}, \mu)$ is called **measure space**.

▶ Remarks:

- A set $A_1 \in \mathscr{A}$ has *zero measure* if there exists $A_2 \in \mathscr{A}$ such that $A_1 \subset A_2$ and $\mu(A_2) = 0$.

- Two sets $A_1, A_2 \subset X$ coincide (mod $\varnothing$) if the symmetric difference $(A_1 \setminus A_2) \cup (A_2 \setminus A_1)$ has zero measure.

- We denote by x a point in $A \in \mathscr{A}$. If a property is valid for all points of $A \subset X$ except for those in a set of measure zero, we say that the property is true for *μ-almost all $x \in A$*.

- Let $(X_i, \mathscr{A}_i, \mu_i)$, $i = 1, \ldots, n$, be measure spaces. The Cartesian product $X := X_1 \times \cdots \times X_n$ has a natural structure of a measure space, whose σ-algebra $\mathscr{A}$ is the smallest σ-algebra of subsets of X containing the subsets of the form

$A_1 \times \cdots \times A_n$, where $A_i \in \mathscr{A}_i$, $i = 1, \ldots, n$. On these subsets the measure μ is defined by

$$\mu(A_1 \times \cdots \times A_n) := \prod_{i=1}^{n} \mu_i(A_i).$$

The space $(X, \mathscr{A}, \mu)$ thus obtained is called *product space* and the measure μ is called *product measure*.

Example 1.1 (*Measure spaces*)

- $(\mathbb{R}, \mathscr{B}(\mathbb{R}), \mu)$ is a measure space with the *Lebesgue measure* $\mu : \mathscr{B}(\mathbb{R}) \rightarrow [0, +\infty]$, which associates with intervals their lengths.

- $(\mathbb{R}^n, \mathscr{B}(\mathbb{R}^n), \mu)$ is a measure space with the *Lebesgue measure* $\mu : \mathscr{B}(\mathbb{R}^n) \rightarrow [0, +\infty]$. It can be proved that μ is the only measure with the property that for any $A := (a_1, b_1) \times \cdots \times (a_n, b_n)$, $(a_i, b_i) \subset \mathbb{R}$, there holds

$$\mu(A) = \prod_{i=1}^{n} (b_i - a_i).$$

- $(X, \mathscr{P}(X), \mu)$, with $X := \{x_1, \ldots, x_N\}$, is a measure space where a measure can be defined by assigning to every element $x_i \in X$ a real number $\mu(x_i) := p_i \geqslant 0$. The measure of the subset $\{x_{i_1}, \ldots, x_{i_k}\} \subset X$ is therefore $p_{i_1} + \cdots + p_{i_k}$. If

$$\sum_{i=1}^{N} p_i = 1,$$

then μ is a *probability measure* and $(X, \mathscr{P}(X), \mu)$ is a *probability space*.

  (a) Let $X := \{0, 1\}$ and $X := \{1, 2, 3, 4, 5, 6\}$ with probabilities $p_1 = p_2 = 1/2$ and $p_1 = \cdots = p_6 = 1/6$ respectively. These spaces can be chosen to represent the probability spaces associated with the toss of a coin and the roll of a die.

  (b) Let $X_1 = \cdots = X_n := \{0, 1\}$ or $\{1, 2, 3, 4, 5, 6\}$ and the measures μ_i coincide with the measure defined in (a). The product space coincides with the space of finite sequences of tosses of a coin or rolls of a die, and the product measure with the probability associated with each sequence.

▶ The theory of Lebesgue measurable functions with its most significant results can be extended to functions $f : X \rightarrow \mathbb{R}$, where $(X, \mathscr{A}, \mu)$ is an arbitrary measure space. For example, $f : X \rightarrow \mathbb{R}$ is said to be measurable if the preimage of each element of $\mathscr{B}(\mathbb{R})$ is measurable, analogous to the situation of continuous functions between topological spaces. Intuitively, a measurable function represents a "measurement" on X.

- Let $f : A \rightarrow [-\infty, +\infty]$, $A \subset X$. The function f is called *measurable* w.r.t. $\mathscr{A}$ if

$$\{x \in A : f(x) < t\} \in \mathscr{A} \qquad \forall t \in \mathbb{R}.$$

One can prove that the above condition is equivalent to $\{x \in A : f(x) \leqslant t\} \in \mathscr{A}$ or $\{x \in A : f(x) > t\} \in \mathscr{A}$ or $\{x \in A : f(x) \geqslant t\} \in \mathscr{A}$.

- The following properties hold:

  (a) The sum and product of two measurable functions are measurable. So is the quotient, if there is no division by zero.

  (b) The composition of measurable functions is measurable.

  (c) The (pointwise) supremum, infimum, limit superior and limit inferior of a sequence of measurable functions are measurable as well.

  (d) The pointwise limit of a sequence of measurable functions is measurable.

- Measurable functions provide a natural context for the theory of integration. We sketch the formal procedure to define the notion of integral on $(X, \mathscr{A}, \mu)$.

  (a) Consider a *finite partition* of X, namely a finite set $\{A_i\}_{1 \leqslant i \leqslant n}$, $A_i \in \mathscr{A}$, such that

  $$\bigcup_{i=1}^{n} A_i = X, \qquad A_i \cap A_j = \emptyset, i \neq j.$$

  (b) Define a *simple function* $g : X \to \mathbb{R}$ by setting

  $$g := \sum_{i=1}^{n} \alpha_i \chi_{A_i}, \qquad \alpha_i \geqslant 0, n \in \mathbb{N},$$

  where χ_{A_i} is the *characteristic function* of A_i:

  $$\chi_{A_i}(x) := \begin{cases} 1 & \text{if } x \in A_i, \\ 0 & \text{if } x \in A_i^c. \end{cases}$$

  (c) Set

  $$\int_X g \, d\mu := \sum_{i=1}^{n} \alpha_i \mu(A_i).$$

  In particular,

  $$\int_X \chi_A \, d\mu := \mu(A) \qquad \forall A \in \mathscr{A}.$$

  (d) If $f : X \to [0, +\infty]$ we set

  $$\int_X f \, d\mu := \sup_{g \in G_f} \int_X g \, d\mu,$$

  where G_f is the set of simple functions g such that $g \leqslant f$. If $f : X \to [-\infty, +\infty]$ we set

  $$\int_X f \, d\mu := \int_X f^+ d\mu - \int_X f^- d\mu, \qquad f^{\pm}(x) := \max(0, \pm f(x)),$$

if at least one of these integrals is finite. In this case f is *μ-summable* on X.
If

$$\int_X |f|\, d\mu < +\infty,$$

we say that f is *μ-integrable* on X.

(e) Let $A \in \mathscr{A}$. Then f is said to be *μ-integrable on A* if the function $f\chi_A$ is μ-integrable on X. We set

$$\int_A f\, d\mu := \int_X f\, \chi_A d\mu.$$

The space of μ-integrable functions on X is denoted by $L^1(X, \mathscr{A}, \mu)$. Similarly, the space of functions f such that $|f|^p, 0 < p < +\infty$, is μ-integrable on X is denoted by $L^p(X, \mathscr{A}, \mu)$.

1.3.2 Probability spaces, random variables and entropy

▶ A probability space is a special measure space. Typically for probability spaces the measure is denoted by P instead of μ.

Definition 1.2

1. A measure space $(X, \mathscr{A}, P)$ is a **probability space** if $P(X) = 1$. The measure $P : \mathscr{A} \to [0,1]$ is called **probability**.

2. A measurable function on $(X, \mathscr{A}, P)$ is called **random variable**. The random variable is **discrete** if its image is a finite or countably infinite set. Otherwise it is called **continuous**.

▶ A more intuitive interpretation of a probability space $(X, \mathscr{A}, P)$ is the following:

- $x \in X$ is an *elementary state*.

- $\mathscr{A}$ is the *family of observable subsets* (or *events*) $A \subset \mathscr{A}$.

- One can usually not decide whether a system is in the particular state $x \in X$, but one can decide whether $x \in A$ or $x \notin A$.

- The measure $P : \mathscr{A} \to [0,1]$ gives a probability to all $A \in \mathscr{A}$. This probability describes how likely it is that the event A occurs.

▶ The following properties hold:

- If $A_1, A_2 \in \mathscr{A}$ then $P(A_1 \setminus A_2) = P(A_1) - P(A_1 \cap A_2)$.

- If $A \in \mathscr{A}$ then $\mathsf{P}(A^c) = 1 - \mathsf{P}(A)$.

- For any sequence $\{A_i\}_{1 \leqslant i \leqslant n}$ of elements of $\mathscr{A}$ we have

$$\mathsf{P}\left(\bigcup_{i=1}^{n} A_i\right) \leqslant \sum_{i=1}^{n} \mathsf{P}(A_i).$$

Equality holds if the elements are disjoint, i.e., $A_i \cap A_j = \emptyset$, $i \neq j$.

Example 1.2 (*A probability space*)

The triple $(\mathbb{R}, \mathscr{B}(\mathbb{R}), \mathsf{P})$ with

$$\mathsf{P}(A) := \frac{1}{\sqrt{2\pi}} \int_A e^{-x^2/2} dx, \qquad A \in \mathscr{B}(\mathbb{R}),$$

is a probability space. The function $1/\sqrt{2\pi}\, e^{-x^2/2}$ is called *probability density function*.

▶ Although a probability space is nothing but a measure space with the measure of the whole space equal to one, probability theory is not merely a subset of measure theory. A distinguishing and fundamental feature of probability theory is the notion of independence of events.

- A sequence of *independent events* $\{A_i\}_{1 \leqslant i \leqslant n}$ is defined by requiring that

$$\mathsf{P}\left(\bigcap_{j=1}^{k} A_{i_j}\right) = \prod_{j=1}^{k} \mathsf{P}(A_{i_j}), \qquad (1.6)$$

for all $\{i_1, \ldots, i_k\} \subseteq \{1, \ldots, n\}$, $1 \leqslant k \leqslant n$. Note that a collection of events $\{A_i\}_{1 \leqslant i \leqslant n}$ may be independent w.r.t. a probability measure P but not w.r.t. another measure P'.

- Given an event $A_1 \in \mathscr{A}$ with $\mathsf{P}(A_1) > 0$ the *conditional probability* of $A_2 \in \mathscr{A}$ w.r.t. A_1 is the number

$$\mathsf{P}(A_2|A_1) := \frac{\mathsf{P}(A_1 \cap A_2)}{\mathsf{P}(A_1)}.$$

- Consider a finite partition $\{A_i\}_{1 \leqslant i \leqslant n}$ of X, where $\mathsf{P}(A_i) > 0$ for all $i = 1, \ldots, n$. Then given an event $A \in \mathscr{A}$ with $\mathsf{P}(A) > 0$ there holds

$$\mathsf{P}(A_i|A) = \frac{\mathsf{P}(A|A_i)\mathsf{P}(A_i)}{\displaystyle\sum_{k=1}^{n} \mathsf{P}(A|A_k)\mathsf{P}(A_k)}.$$

▶ We now introduce the concept of entropy on a probability space.

- Let $(X, \mathscr{A}, P)$ be a probability space with

$$X := \bigcup_{i=1}^{n} A_i, \qquad A_i \cap A_j = \varnothing, \ i \neq j,$$

and

$$P(A_i) := p_i \in [0,1], \qquad i = 1, \ldots, n.$$

Concretely, the finite partition X can be interpreted as an experiment with n possible mutually exclusive outcomes $A_1, \ldots, A_n$ (for example the toss of a coin, $n = 2$, or the roll of a die, $n = 6$), where each outcome A_i happens with probability p_i.

- Let $\triangle^{(n)}$ be the $(n-1)$-dimensional simplex of $\mathbb{R}^n$ defined by

$$\triangle^{(n)} := \left\{ (p_1, \ldots, p_n) \in [0,1]^n : \sum_{i=1}^{n} p_i = 1 \right\}.$$

- A one-parameter family of functions $H^{(n)} \in C(\triangle^{(n)}, [0, +\infty])$ is called *entropy* if:

    1. For all $i, j \in \{1, \ldots, n\}$ we have

    $$H^{(n)}(p_1, \ldots, p_i, \ldots, p_j, \ldots, p_n) = H^{(n)}(p_1, \ldots, p_j, \ldots, p_i, \ldots, p_n).$$

    2. $H^{(n)}(1, 0, \ldots, 0) = 0$.

    3. $H^{(n)}(0, p_2, \ldots, p_n) = H^{(n-1)}(p_2, \ldots, p_n)$ for all $n \geqslant 2$ and $(p_2, \ldots, p_n) \in \triangle^{(n-1)}$.

    4. $H^{(n)}(p_1, \ldots, p_n) \leqslant H^{(n)}(1/n, \ldots, 1/n)$ for all $(p_1, \ldots, p_n) \in \triangle^{(n)}$. Equality holds if and only if $p_i = 1/n$ for all $i = 1, \ldots, n$.

    5. Let $(p_{11}, \ldots, p_{1\ell}, p_{21}, \ldots, p_{2\ell}, \ldots, p_{n1}, \ldots, p_{n\ell}) \in \triangle^{(n\ell)}$. Then we have

    $$H^{(n\ell)}(p_{11}, \ldots, p_{n\ell}) = H^{(n)}(p_1, \ldots, p_n) + \sum_{i=1}^{n} p_i H^{(\ell)}\left(\frac{p_{i1}}{p_i}, \ldots, \frac{p_{i\ell}}{p_i} \right),$$

    for all $(p_1, \ldots, p_n) \in \triangle^{(n)}$.

- The above definition describes the five properties which must hold for a function measuring the *uncertainty* of the prediction of an outcome of the experiment (equivalently, the *information* acquired from the execution of the experiment). Here is the meaning of the five above listed properties:

1. Symmetry of the functions $H^{(n)}$.
2. Absence of uncertainty of a certain event.
3. No information is gained by impossible outcomes.
4. Maximal uncertainty is attained when all outcomes are equally probable.
5. Behavior of the entropy when distinct experiments are compared. Consider a second experiment with possible outcomes $B_1, \ldots, B_\ell$ (i.e., another finite partition of $(X, \mathscr{A}, \mathsf{P})$). Let p_{ij} be the probability of A_i and B_j together. The conditional probability of B_j w.r.t. A_i is $\mathsf{P}(B_j|A_i) = p_{ij}/p_i$. The uncertainty in the prediction of the outcome of the second experiment when the outcome of the first one is given by A_i is measured by $H^{(\ell)}(p_{i1}/p_i, \ldots, p_{i\ell}/p_i)$. From this fact follows the requirement that the fifth property be satisfied.

- Remarkably, it can be proved that given $(p_1, \ldots, p_n) \in \triangle^{(n)}$ the function

$$H^{(n)}(p_1, \ldots, p_n) := -\sum_{i=1}^{n} p_i \log p_i, \tag{1.7}$$

with the convention $0 \log 0 = 0$, is, up to a constant positive factor, the only function satisfying the five listed properties.

1.3.3 Ergodic dynamical systems

▶ Ergodic theory investigates dynamical systems, defined as (semi)group actions on sets, which preserve a probability measure.

▶ Let $(X, \mathscr{A}, \mathsf{P})$ be a probability space. We interpret X as the *phase space* of a dynamical system defined in terms of (iterations of) a map $\Phi : X \to X$.

- Φ is *measurable* if $\Phi^{-1}(A) := \{x \in X : \Phi(x) \in A\} \in \mathscr{A}$ for all $A \in \mathscr{A}$.

- Φ is *non-singular* if it is measurable and $\mathsf{P}(\Phi^{-1}(A)) = 0$ for all $A \in \mathscr{A}$ such that $\mathsf{P}(A) = 0$.

- Φ is *measure-preserving* if it is measurable, non-singular and $\mathsf{P}(\Phi^{-1}(A)) = \mathsf{P}(A)$ for all $A \in \mathscr{A}$.

- If Φ is measure-preserving then $(X, \mathscr{A}, \mathsf{P}, \Phi)$ is a *measurable dynamical system*.

  (a) An *orbit* of a point $x \in X$ is the infinite sequence of points

  $$\left(\Phi^j(x)\right)_{j \in \mathbb{N}} := \left(x, \Phi(x), \Phi^2(x), \ldots, \Phi^{n+1}(x), \ldots\right),$$

  where $\Phi^{n+1}(x) := \Phi(\Phi^n(x))$. This represents one entire history of the system.

(b) The famous "Poincaré recurrence Theorem" admits a generalization valid for measurable dynamical systems: *Let $(X, \mathscr{A}, \mathsf{P}, \Phi)$ be a measurable dynamical system. For any $A \in \mathscr{A}$ the subset B of all points $x \in A$ such that $\Phi^n(x) \in A$, for infinitely many values of $n \in \mathbb{N}$, belongs to $\mathscr{A}$ and $\mathsf{P}(A) = \mathsf{P}(B)$.*

▶ A first general problem is to determine all measures which are invariant under the action of a given map. We sketch a characterization of this problem in a particular case: we consider the set of probability measures on $([0,1], \mathscr{B}([0,1]))$ which can be written as

$$\mathsf{P}(A) := \int_A \rho \, d\mu, \qquad A \in \mathscr{B}([0,1]),$$

where ρ is a positive-valued function, called *probability density function*, and μ is the Lebesgue measure. In other words, we simply write

$$\mathsf{P}(A) := \int_A \rho(x) \, dx,$$

and we want to determine all such P that are invariant under a given differentiable and piecewise monotone measurable map $\Phi : [0,1] \to [0,1]$.

- The invariance of the measure is expressed by the condition

$$\mathsf{P}(A) = \int_A \rho(x) \, dx = \int_{\Phi^{-1}(A)} \rho(x) \, dx = \mathsf{P}\left(\Phi^{-1}(A)\right) \qquad \forall A \in \mathscr{A}. \quad (1.8)$$

- There exists a finite or countable decomposition of the interval $[0,1]$ into intervals $[a_i, a_{i+1}]$, $i \in J := \{1, \ldots, k\}$, $k \in \mathbb{N}$, on which Φ is differentiable and monotone. Denote by Φ_i^{-1} the well-defined inverse of Φ on each of these subintervals.

- Define $A := [0, x]$, $x \leqslant 1$. Condition (1.8) becomes

$$\int_0^x \rho(s) \, ds = \sum_{i \in J} \int_{\Phi_i^{-1}([0,x])} \rho(s) \, ds. \quad (1.9)$$

Differentiating (1.9) w.r.t. x, we obtain

$$\rho(x) = \sum_{i \in J_x} \frac{\rho\left(\Phi_i^{-1}(x)\right)}{\left|\Phi'\left(\Phi_i^{-1}(x)\right)\right|}, \quad (1.10)$$

where J_x indicates the subset of J corresponding to indices i such that $\Phi_i^{-1}(x) \neq \varnothing$. Here $\Phi'(x) = d\Phi/dx$.

- Equation (1.10) is a necessary and sufficient condition for P to be invariant w.r.t. Φ.

Example 1.3 (*Ulam-Von Neumann map*)

Consider the probability space $([0,1], \mathscr{B}([0,1]), P)$, where

$$P(A) := \int_A \rho(x)\, dx, \qquad A \in \mathscr{B}([0,1]),$$

with

$$\rho(x) := \frac{1}{\pi} \frac{1}{\sqrt{x(1-x)}}.$$

We can verify that P is an invariant measure under the action of the map

$$\Phi(x) := 4\,x(1-x).$$

Note that $\Phi'(x) = 4 - 8\,x$. To every point $x \in [0,1]$ there correspond two preimages

$$\Phi_1^{-1}(x) = \frac{1}{2}\left(1 - \sqrt{1-x}\right) \in [0,1/2],$$

and

$$\Phi_2^{-1}(x) = \frac{1}{2}\left(1 + \sqrt{1-x}\right) \in [1/2,1].$$

Condition (1.10) is fulfilled:

$$\frac{\rho\left(\Phi_1^{-1}(x)\right)}{\left|4 - 8\,\Phi_1^{-1}(x)\right|} + \frac{\rho\left(\Phi_2^{-1}(x)\right)}{\left|4 - 8\,\Phi_2^{-1}(x)\right|} = \frac{1}{\sqrt{x(1-x)}}.$$

▶ A second general problem is to understand how often the orbit of a given point x of a measurable dynamical system $(X, \mathscr{A}, P, \Phi)$ visits a prescribed measurable set $A \in \mathscr{A}$. Here we list the most important notions and facts.

- Let χ_A be the characteristic function of A. For every $n \in \mathbb{N}$, the *number of visits* (within the time n) of A by the orbit of x is the number

$$K(x, A, n) := \sum_{j=0}^{n-1} \chi_A\left(\Phi^j(x)\right).$$

- The *frequency of visits* of A by the orbit of x is the limit

$$\nu(x, A) := \lim_{n \to +\infty} \frac{1}{n} K(x, A, n).$$

Remarkably, it can be proven that for P-almost every $x \in X$ the frequency of visits $\nu(x, A)$ does exist.

- A measurable dynamical system is *ergodic* if for every choice of $A \in \mathscr{A}$ there holds

$$\nu(x, A) = P(A)$$

for P-almost every $x \in X$. In this case P is said to be an *ergodic measure* w.r.t. Φ and $(X, \mathscr{A}, P, \Phi)$ is an *ergodic dynamical system*.

- If instead of the characteristic function of a set we consider arbitrary integrable functions we get the famous "Birkhoff Theorem": *The time average of $f \in L^1(X, \mathscr{A}, \mathrm{P})$ along the orbit of $x \in X$, defined by*

$$\langle f(x) \rangle_\infty := \lim_{n \to +\infty} \frac{1}{n} \sum_{j=0}^{n-1} f\left(\Phi^j(x)\right), \tag{1.11}$$

  exists for P*-almost every* $x \in X$.

▶ Remarks:

- Note that $\langle f(\Phi(x)) \rangle_\infty = \langle f(x) \rangle_\infty$ for P-almost every $x \in X$. Hence the time average depends on the orbit and not on the initial point chosen along the orbit.

- The measure P is Φ-invariant. Therefore we define the *phase space average* (or *expectation value*) of f as

$$\langle f(x) \rangle_\mathrm{P} := \int_X f(x) \, d\mathrm{P} = \int_X f(\Phi(x)) \, d\mathrm{P}.$$

- An application of Lebesgue dominated convergence to (1.11) implies

$$\langle f(x) \rangle_\mathrm{P} = \int_X \langle f(x) \rangle_\infty \, d\mathrm{P} = \langle \langle f(x) \rangle_\infty \rangle_\mathrm{P},$$

  which means that f and its time average have the same expectation value. Then a possible characterization of the ergodicity of $(X, \mathscr{A}, \mathrm{P}, \Phi)$ is that for every $f \in L^1(X, \mathscr{A}, \mathrm{P})$ there holds

$$\langle f(x) \rangle_\mathrm{P} = \langle f(x) \rangle_\infty$$

  for P-almost every $x \in X$.

▶ Ergodicity gives a qualitative indication of the degree of randomness of a measurable dynamical system. An indication of quantitative type is given by the notion of entropy, which generalizes the notion of entropy given for a probability space. Nevertheless the formal definition of entropy of a measurable dynamical system will not be covered in this course. Roughly speaking, this quantity allows one to distinguish between systems in terms of the "predictability" of their observables.

1.4 Exercises

Ch1.E1 Consider a probability space and an event $A := A_1 \cup A_2$, where A_1 and A_2 are two disjoint events with $P(A_1) := p \in [0,1]$ and $P(A_2) := 1 - p$. Find the entropy $H(p)$.

Ch1.E2 Consider an experiment for which there are only two possible outcomes A_1 and A_2 to an observation made on it. Let their probabilities of occurrence be $P(A_1)$ and $P(A_2)$.

Suppose that we make three independent observations of the experiment. Determine the probability of two occurrences of the outcome A_1 and one occurrence of the outcome A_2, with no regard to the order of these occurrences.

Ch1.E3 Consider N (not necessarily disjoint) events $\{A_i\}_{1 \leqslant i \leqslant N}$. Prove that

$$P\left(\bigcup_{i=1}^{N} A_i\right) \leqslant \sum_{i=1}^{N} P(A_i).$$

Ch1.E4 Prove that if A and B are two independent events, i.e., $P(A \cap B) = P(A)P(B)$, then also the complementary events A^c and B^c are independent.

Ch1.E5 Suppose that there are 25 students in a classroom. What is the probability that at least two classmates have the same birthday?

Ch1.E6 An airplane needs at least half of its engines to safely complete its mission. If each engine independently works fine with probability $p \in [0,1]$, for what values of p is a three-engine plane safer than a five-engine plane?

Ch1.E7 Consider an experiment with two possible mutually exclusive outcomes A_1 and A_2, where the outcome A_1 can be observed with probability $P(A_1) = p \in [0,1]$.

(a) Suppose we carry out the experiment N times so that the observations are statistically independent and compute the probability that we observe A_1 exactly k times.

(b) Compute the probability that in the above sequence of independent experiments we observe A_1 for the first time at the k-th experiment.

(c) Compute the probability that we eventually observe A_1 when N goes to infinity.

2

Introduction to Kinetic Theory of Gases

2.1 Introduction

▶ Around the year 1870 Boltzmann proposed that macroscopic laws of thermodynamics should be derivable from mechanical first principles on the basis of atomistic theory of matter.

▶ The object of investigation is a theoretical *gas of particles* so defined (*ideal gas*):

1. *Hard spheres.* N (say $N \approx 6.02 \times 10^{23}$, *Avogadro number*) identical hard spheres (radius r, mass m) without internal structure contained in a bounded region $\Lambda \subset \mathbb{R}^3$, $\mathrm{Vol}(\Lambda) = V$. The volume V is not necessarily constant in time.

2. *Strong dilution.* Let $n := N/V$ be the number of particles per unit volume. We assume that $n\,r^3 \ll 1$ and therefore the probability that two particles are at a distance of order r (hence colliding) is small.

3. *Interactions through perfectly elastic binary collisions.* The random motion of particles obeys Newton's laws. All collisions between particles and with the boundary $\partial \Lambda$ are perfectly elastic, i.e., non-dissipative. Further, we exclude all situations where three or more particles collide at the same time. This last assumption is reasonable if assumption 2 holds true, because the mean free path of a particle (the average distance between two consecutive collisions) is then much larger than the average diameter of particles.

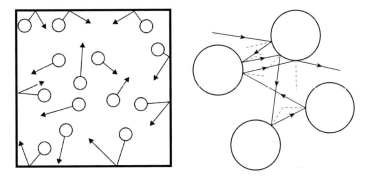

Fig. 2.1. Ideal gas ([FaMa]).

▶ Remarks:

- The motion of particles of an ideal gas can be influenced by an external (con-servative) force, which, for example, leads to a variation of V. If there are no external fields acting on the system the gas is called *free ideal gas* (or *perfect gas*).

- To assume that particles are hard spheres without internal structure is a strong and restrictive hypothesis. This assumption says that rotational modes are not allowed.

- Interactions between particles give rise to observable macroscopic quantities, as for instance the pressure P of the gas, which is originated by the collisions of particles with the boundary $\partial \Lambda$. The thermal variables, as for instance temper-ature and entropy, should emerge as effective quantities describing the macro-scopic features of the microscopic dynamical state of the gas that would other-wise be disregarded.

- The *free ideal gas law* is the equation of state of a free ideal gas at equilibrium. It was obtained empirically in the 19th century and it reads

$$PV = N\kappa T, \qquad (2.1)$$

where $\kappa > 0$ is a universal constant called *Boltzmann constant*.

- At normal conditions such as standard temperature and pressure, most real molecular gases behave qualitatively like an ideal gas and satisfy (2.1). Gener-ally, a gas behaves more like an ideal gas at high temperatures and low pres-sures.

2.2   The Boltzmann kinetic theory of gases

▶ The task of the Boltzmann kinetic theory is to study the time evolution of a gas of particles, not necessarily of ideal type, and the achievement of an equilibrium state, for which collisions between particles play a crucial role. A remarkable goal of kinetic theory of gases was to derive the free ideal gas law (2.1) on the basis of a microscopic statistical analysis of the model.

▶ Our interest goes to the investigation of an ideal gas (with V constant), with and without external forces. Here we list the two main mathematical/physical object of interest.

1. *Phase space.* According to a canonical Hamiltonian description of motion we use a six-dimensional phase space $\Delta := \Lambda \times \mathbb{R}^3$ whose (time-dependent, $t \subset \mathbb{R}$) generalized canonical coordinates are denoted by (q, p). The dynamical state of a single particle of the gas at time t is completely determined by a point in Δ. In other words, at every time t, the kinematic state of the gas is completely defined in terms of N points in Δ.

2. *Probability distribution function on the phase space.* Consider in Δ a cell Δ_0 centered at (q, p).

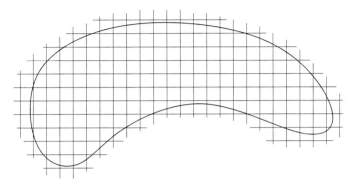

Fig. 2.2. Decomposition of Δ into cells ([Ue]).

We define $\nu(\Delta_0, t)$ to be the number of representative points in Δ_0 at time t. For typical values of the ratio N/V (which are compatible with the assumption that the gas is ideal), the ratio $\nu(\Delta_0, t)/\text{Vol}(\Delta_0)$ stabilizes, as the diameter of the cell becomes sufficiently small, to a value depending on the center of the cell, (q, p), and on the time t considered. This value defines a positive definite function $\varrho : \mathbb{R} \times \Delta \to \mathbb{R}_+$, called *distribution function* of the gas.

(a) The number of particles $\nu(\Delta_0, t)$, whose kinematic state at time t is described by a point in $\Delta_0 \subset \Delta$, is given by the integral

$$\nu(\Delta_0, t) := \int_{\Delta_0} \varrho(q, p, t) \, dq \, dp. \tag{2.2}$$

In a probabilistic language, $\varrho(q, p, t) \, dq \, dp$ expresses the product between N and the probability that a particle is found in the infinitesimal phase space volume $dq \, dp$ centered at (q, p) at time t. By construction we have

$$N = \int_{\Delta} \varrho(q, p, t) \, dq \, dp. \tag{2.3}$$

All integrals are assumed to be convergent. Note that to get a *probability distribution function* we can simply set

$$\widetilde{\varrho}(q, p, t) := \frac{\varrho(q, p, t)}{N},$$

whose integral over Δ is now normalized to one.

(b) If the spatial distribution of the particles is uniform, a condition which is compatible with absence of external forces, the distribution function is independent of $q \in \Lambda$ and integration w.r.t. q in (2.3) simply leads to the factorization of the volume V. In this case, we obtain the following expression for the number of particles per unit volume, the so called *density of particles* (at time t):

$$n(t) := \frac{N}{V} = \int_{\mathbb{R}^3} \varrho(p, t) \, dp.$$

(c) The main task is to derive macroscopic observable quantities from the distribution function ϱ. Thus, if $f : \Delta \to \mathbb{R}$ is a function representing a measurable physical quantity we are interested in, then its *phase space average* at time t w.r.t. ϱ is given by

$$\langle f(q, p) \rangle_\varrho (t) := \frac{\int_\Delta f(q, p) \, \varrho(q, p, t) \, dq \, dp}{\int_\Delta \varrho(q, p, t) \, dq \, dp},$$

where it is assumed that the integral converges.

▶ It is natural to identify the above formalism with the formalism of measure spaces defined in Chapter 1. Indeed, we can write

$$(X, \mathscr{A}, \mu) \equiv (\Delta, \mathscr{B}(\Delta), \nu),$$

where ν is defined by (2.2) for any $\Delta_0 \in \mathscr{B}(\Delta)$.

▶ The aim is to find the structure of ϱ and deduce the set of differential equations that ϱ obeys. In particular, those distribution functions ϱ which correspond to an equilibrium state should allow one to derive the thermodynamics of the gas.

- *Equilibrium statistical mechanics.* J.C. Maxwell (1831-1879) was the first to look at gases from a probabilistic point of view. He considered an ideal gas without external forces, i.e., a free ideal gas, and computed the *equilibrium probability distribution function* on the basis of the "Central limit Theorem" of probability theory. Here the term "equilibrium" refers to the fact that the ideal gas is in *thermodynamic equilibrium,* a state of balance corresponding to thermal equilibrium, mechanical equilibrium, radiative equilibrium, and chemical equilibrium. Thermodynamic equilibrium is the unique stable stationary state that is approached or eventually reached as the system interacts with its surroundings over a long time.

  (a) This analysis leads to an equilibrium probability distribution function for the momentum p of the particles, say $\varrho_0(p)$. It is called *Maxwell distribution* and it allows one to obtain the free ideal gas law (2.1).

(b) A generalization of the Maxwell distribution, valid for an ideal gas subject to external conservative forces, leads to a distribution function which depends both on q and p, say $\varrho(q, p)$. It is called *Boltzmann-Maxwell distribution*.

- *Non-equilibrium statistical mechanics.* L. Boltzmann (1844-1906) was more interested in understanding how the thermodynamic equilibrium, corresponding to the Boltzmann-Maxwell distribution $\varrho(q, p)$, can be achieved in terms of collisions between particles and how this process can be related to time variations of a *non-equilibrium distribution function* $\varrho(q, p, t)$ of the gas:

$$\frac{d\varrho}{dt} = \frac{\partial \varrho}{\partial t} + \Big\langle \text{grad}_q \varrho(q, p, t), \dot{q} \Big\rangle + \Big\langle \text{grad}_p \varrho(q, p, t), \dot{p} \Big\rangle . \tag{2.4}$$

We introduced the *"dot" notation* for time-derivatives: $\dot{q} \equiv dq/dt$, $\dot{p} \equiv dp/dt$.

(a) It is natural to expect that $\varrho(q, p, t)$ converges to the Boltzmann-Maxwell distribution $\varrho(q, p)$ when the thermodynamic equilibrium is achieved.

(b) Equation (2.4) is considered in the framework of Newtonian mechanics. Under some quite relaxed assumptions it describes a mechanical process which is *time-recurrent* (i.e., the system will, after a sufficiently long but finite time, return to a state very close to the initial state) and *time-reversible* (i.e., dynamics is invariant under the change $t \mapsto -t$). Note that if we assume that the particles move under the influence of some external potential energy $\mathscr{U} : \Lambda \to \mathbb{R}$, then we can write

$$\begin{aligned} \frac{d\varrho}{dt} &= \frac{\partial \varrho}{\partial t} + \frac{1}{m} \Big\langle \text{grad}_q \varrho(q, p, t), p \Big\rangle \\ &\quad - \Big\langle \text{grad}_p \varrho(q, p, t), \text{grad}_q \mathscr{U}(q) \Big\rangle, \end{aligned} \tag{2.5}$$

where we used $(\dot{q}, \dot{p}) = (p/m, - \text{grad}_q \mathscr{U}(q))$.

2.2.1 Derivation of the Boltzmann transport equation

▶ We present here Boltzmann's approach to derive the evolution equation governing the time evolution of the probability distribution function. We will end up with a remarkable integral-differential equation, called *Boltzmann transport equation*. We will be interested not in its time-dependent general solutions, but rather in its stationary solutions, namely those time-independent solutions corresponding to thermodynamic equilibrium.

▶ We already defined our ideal gas, see conditions 1,2 and 3 in Section 2.1. We now add a fourth crucial (and controversial) hypothesis:

4. *Stosszahlansatz* (*collision number hypothesis*). The distribution function of a pair of colliding particles, hence the probability that at time t we can determine a binary collision at position q between two particles with momenta p_1 and p_2, is proportional to the product $\varrho(q, p_1, t)\, \varrho(q, p_2, t)$. This is equivalent to postulating weak correlation between the motion of the two colliding particles before the collision, i.e., independence of the probability densities of colliding particles (see (1.6)). A posteriori, we will see that such hypothesis breaks the time-reversibility of the process.

Hereafter, in the context of kinetic theory, we call ideal gas a gas satisfying hypotheses 1-4 and not only 1-3.

▶ On the basis of our hypotheses 1-4 we can now derive the form of the evolution equation of the probability distribution $\varrho(q, p, t)$ of the ideal gas:

• Consider two colliding particles, say 1 and 2. We denote by $p_i \in \mathbb{R}^3$, $i = 1, 2$, the momenta before collision and by $\tilde{p}_i \in \mathbb{R}^3$, $i = 1, 2$, the momenta after collision. The perfect elasticity of the collision implies the *conservation of total momentum* and the *conservation of mechanical energy*:

$$p_1 + p_2 = \tilde{p}_1 + \tilde{p}_2, \tag{2.6}$$
$$\|p_1\|^2 + \|p_2\|^2 = \|\tilde{p}_1\|^2 + \|\tilde{p}_2\|^2. \tag{2.7}$$

• If we call $\tau(p_1, p_2, \tilde{p}_1, \tilde{p}_2)$ the probability of the transition $(p_1, p_2) \to (\tilde{p}_1, \tilde{p}_2)$ we obtain a positive definite function $\tau = \tau(p_1, p_2, \tilde{p}_1, \tilde{p}_2)$, called *transition kernel*, which satisfies the following properties:

  1. τ is symmetric w.r.t. $(p_1, p_2) \leftrightarrow (\tilde{p}_1, \tilde{p}_2)$ because the inverse transition has the same probability, due to the time-reversibility of Newton equations.

  2. τ is symmetric w.r.t. $p_1 \leftrightarrow p_2$ and $\tilde{p}_1 \leftrightarrow \tilde{p}_2$ because particles are identical.

For reasons of isotropy, it is reasonable to assume that τ depends on the modulus of the relative momentum of the colliding particles, in addition to the angular coordinates of the collision.

• Introduce the functions

$$\varrho_i := \varrho(q, p_i, t), \qquad \tilde{\varrho}_i := \varrho(q, \tilde{p}_i, t), \qquad i = 1, 2. \tag{2.8}$$

Consider the function ϱ_1. This is the probability distribution function for particles with momentum p_1. We are interested in determining the time evolution of ϱ_1 in terms of the transition kernel τ and the functions (2.8), taking into account the validity of conservation laws (2.6) and (2.7).

- Note that $d\varrho_1/dt$ is the sum of a negative term, due to the transitions $(p_1, p_2) \rightarrow (\widetilde{p}_1, \widetilde{p}_2)$ for any p_2, and of a positive term due to the inverse transition. For fixed p_1, we must consider all possible vectors p_2 and all the possible pairs $(\widetilde{p}_1, \widetilde{p}_2)$ that are compatible with the conservation laws (2.6) and (2.7).

- The *Stosszahlansatz* implies that the frequency of the transitions $(p_1, p_2) \rightarrow (\widetilde{p}_1, \widetilde{p}_2)$ and the frequency of the inverse ones are proportional to the products $\varrho_1 \varrho_2$ and $\widetilde{\varrho}_1 \widetilde{\varrho}_2$.

- The transition kernel weighs products $\varrho_1 \varrho_2$ and $\widetilde{\varrho}_1 \widetilde{\varrho}_2$ to obtain the corresponding frequencies. Hence at every point q, for fixed p_1 and p_2, the frequency of the collisions that make a particle leave the class described by the function ϱ_1 is $\tau(p_1, p_2, \widetilde{p}_1, \widetilde{p}_2) \varrho_1 \varrho_2$ while the frequency of the collisions that join this class is $\tau(p_1, p_2, \widetilde{p}_1, \widetilde{p}_2) \widetilde{\varrho}_1 \widetilde{\varrho}_2$. For notational convenience we introduce the function

$$\delta(q, p_1, p_2, \widetilde{p}_1, \widetilde{p}_2, t) := \varrho(q, \widetilde{p}_1, t) \varrho(q, \widetilde{p}_2, t) - \varrho(q, p_1, t) \varrho(q, p_2, t).$$

- To obtain the collision term that equates with $d\varrho_1/dt$ given by (2.5) we must therefore integrate the expression $\tau(p_1, p_2, \widetilde{p}_1, \widetilde{p}_2) \delta(q, p_1, p_2, \widetilde{p}_1, \widetilde{p}_2, t)$ over all momenta p_2 and over the regular two-dimensional submanifold of $\mathbb{R}^6$ spanned by all pairs $(\widetilde{p}_1, \widetilde{p}_2)$ subject to the constraints (2.6) and (2.7), where the invariants are fixed by the values p_1, p_2. Let us denote this submanifold by $\Sigma_{p_{\text{tot}}, E_{\text{tot}}}$:

$$\Sigma_{p_{\text{tot}}, E_{\text{tot}}} := \left\{ (\widetilde{p}_1, \widetilde{p}_2) \in \mathbb{R}^6 \ : \ \widetilde{p}_1 + \widetilde{p}_2 = p_{\text{tot}}, \ \frac{1}{2m}\left(\|\widetilde{p}_1\|^2 + \|\widetilde{p}_2\|^2 \right) = E_{\text{tot}} \right\},$$

with p_{tot} fixed by $p_1 + p_2$ and E_{tot} fixed by $\left(\|p_1\|^2 + \|p_2\|^2 \right) / (2m)$.

▶ Therefore we obtained the non-equilibrium transport equation we were looking for. The (non-equilibrium) probability distribution function ϱ_1 obeys the following integral-differential equation:

$$\frac{d\varrho_1}{dt} = \int_{\mathbb{R}^3} dp_2 \int_{\Sigma_{p_{\text{tot}}, E_{\text{tot}}}} \tau(p_1, p_2, \widetilde{p}_1, \widetilde{p}_2) \, \delta(q, p_1, p_2, \widetilde{p}_1, \widetilde{p}_2, t) \, d\widetilde{p}_1 d\widetilde{p}_2. \qquad (2.9)$$

This equation is called *Boltzmann transport equation*.

2.2.2   Equilibrium solutions of the Boltzmann transport equation

▶ The thermodynamic equilibrium corresponds to those time-independent distribution functions such that

$$\int_{\mathbb{R}^3} dp_2 \int_{\Sigma_{p_{\text{tot}}, E_{\text{tot}}}} \tau(p_1, p_2, \widetilde{p}_1, \widetilde{p}_2) \, \delta(q, p_1, p_2, \widetilde{p}_1, \widetilde{p}_2) \, d\widetilde{p}_1 \, d\widetilde{p}_2 = 0.$$

Such equilibrium solution functions describe how canonical coordinates $(q, p) \in \Delta$ are distributed when the gas reaches its equilibrium.

▶ Before deriving equilibrium solutions of (2.9) we need a technical lemma which will be used many times. Its proof can be found in any Analysis textbook.

Lemma 2.1

Let $n \in \mathbb{N}$, $a > 0$ and $x \in \mathbb{R}$.

1. *Given*

$$\mu_n := \int_{-\infty}^{+\infty} x^n e^{-a x^2} dx,$$

   there holds

$$\mu_0 = \sqrt{\frac{\pi}{a}}, \qquad \mu_{2n+1} = 0, \qquad \mu_{2n} = (2n-1)!! \sqrt{\frac{\pi}{a}} (2a)^{-n}.$$

2. *There holds*

$$\int_0^{+\infty} x^n e^{-a x^2} dx = \frac{a^{-(n+1)/2}}{2} \Gamma\left(\frac{n+1}{2}\right),$$

   *where Γ is the **Euler Γ-function** defined by*

$$\Gamma(x) := \int_0^{+\infty} t^{x-1} e^{-t} dt, \qquad x > 0.$$

3. *The following formulas hold true:*

   (a) *$\Gamma(x+1) = x \Gamma(x)$ and $\Gamma(n+1) = n!$.*
   (b) *$\Gamma(n+1/2) = (2n)! \sqrt{\pi} / (4^n n!)$.*
   (c) *Assume $x \gg 1$. Then:*

$$\Gamma(x) \approx \sqrt{2\pi} \, x^{x+1/2} e^{-x}, \qquad \log \Gamma(x) \approx x \log x - x. \qquad (2.10)$$

   Assume $n \gg 1$. Then:

$$n! \approx \sqrt{2\pi n} \left(\frac{n}{e}\right)^n, \qquad \log n! \approx n \log n - n. \qquad (2.11)$$

   *The above formulas are called **Stirling approximations**.*

No Proof.

▶ We are now ready to derive the equilibrium solutions of (2.9), namely the so called *Boltzmann-Maxwell distribution function*.

Theorem 2.1 (*Boltzmann–Maxwell*)

Consider an ideal gas and let $\mathcal{U} : \Lambda \to \mathbb{R}$ be an external potential energy influencing the motion of particles constituting the gas. The equilibrium solutions of (2.9) are

$$\varrho(q, p) = \varrho_0(p) \left(\frac{1}{V} \int_\Lambda e^{-3\,\mathcal{U}(q)/(2\,\varepsilon)} \mathrm{d}q \right)^{-1} e^{-3\,\mathcal{U}(q)/(2\,\varepsilon)}, \qquad (2.12)$$

where

$$\varrho_0(p) := n \left(\frac{3}{4\,\pi\,\varepsilon\,m} \right)^{3/2} e^{-3\|p\|^2/(4\,m\,\varepsilon)}. \qquad (2.13)$$

Here $n := N/V$ is the number of particles per unit volume and ε is the average kinetic energy of a particle:

$$\varepsilon := \frac{1}{2\,m} \left\langle \|p\|^2 \right\rangle_{\varrho_0}.$$

Proof. We proceed by steps.

- Assume that $\mathcal{U}$ is zero. This hypothesis implies that ϱ is spatially uniform, i.e., it does not depend on q. Hence we set $\varrho(q, p) = \varrho_0(p)$. From (2.9) we see that a sufficient condition for $\varrho_0(p)$ to be stationary is that it satisfies the functional equation

$$\delta\,(p_1, p_2, \tilde{p}_1, \tilde{p}_2) = 0,$$

  namely

$$\varrho_0(p_1)\,\varrho_0(p_2) = \varrho_0(\tilde{p}_1)\,\varrho_0(\tilde{p}_2), \qquad (2.14)$$

  for every pair (p_1, p_2), $(\tilde{p}_1, \tilde{p}_2)$ satisfying (2.6) and (2.7). This means that the product $\varrho_0(p_1)\,\varrho_0(p_2)$ must depend only on the conserved total energy

$$E_{\text{tot}} := \frac{1}{2\,m} \left(\|p_1\|^2 + \|p_2\|^2 \right),$$

  and on the conserved total momentum

$$p_{\text{tot}} := p_1 + p_2.$$

- A possible choice for ϱ_0 satisfying (2.14) such that $\varrho_0(p) \to 0$ for $\|p\| \to +\infty$ is

$$\varrho_0(p) = C\,e^{-A\|p-p_0\|^2},$$

  where $A, C > 0$ are two constants to be determined and $p_0 \in \mathbb{R}^3$ is an arbitrary vector. Note that

$$\varrho_0(p_1)\,\varrho_0(p_2) = C^2\,e^{-A(\|p_1-p_0\|^2+\|p_2-p_0\|^2)},$$

where

$$\|p_1 - p_0\|^2 + \|p_2 - p_0\|^2 \;=\; \|p_1\|^2 + \|p_2\|^2 - 2 \langle p_1 + p_2, p_0 \rangle + 2 \|p_0\|^2$$
$$=\; 2 m \, E_{tot} - 2 \langle p_{tot}, p_0 \rangle + 2 \|p_0\|^2,$$

so that all our constraints are fulfilled. Without any loss of generality we can fix $p_0 = 0$.

- The normalizing condition of the distribution function fixes the constant C in terms of A. Indeed, by using spherical coordinates and using Lemma 2.1, we get

$$
\begin{aligned}
n \;:=\; \frac{N}{V} &= \int_{\mathbb{R}^3} \varrho_0(p) \, dp \\
&= 4 \pi C \int_0^{+\infty} \|p\|^2 e^{-A\|p\|^2} d\|p\| \\
&= 4 \pi C \, \frac{A^{-3/2}}{2} \, \Gamma\left(\frac{3}{2}\right) \\
&= C \left(\frac{\pi}{A}\right)^{3/2},
\end{aligned}
$$

where we used Lemma 2.1 (note that $\Gamma(3/2) = \sqrt{\pi}/2$). Therefore we have

$$C = n \left(\frac{A}{\pi}\right)^{3/2}. \tag{2.15}$$

- The constant A can be related to the average kinetic energy ε of a particle:

$$
\begin{aligned}
\varepsilon := \frac{1}{2m} \left\langle \|p\|^2 \right\rangle_{\varrho_0} &:= \frac{1}{2m} \frac{\int_{\mathbb{R}^3} \|p\|^2 \varrho_0(p) \, dp}{\int_{\mathbb{R}^3} \varrho_0(p) \, dp} \\
&= \frac{2\pi}{m} \left(\frac{A}{\pi}\right)^{3/2} \int_0^{+\infty} \|p\|^4 e^{-A\|p\|^2} d\|p\| \\
&= \frac{2\pi}{m} \left(\frac{A}{\pi}\right)^{3/2} \frac{A^{-5/2}}{2} \, \Gamma\left(\frac{5}{2}\right) \\
&= \frac{3}{4 A m},
\end{aligned}
$$

where we used Lemma 2.1 (note that $\Gamma(5/2) = 3\sqrt{\pi}/4$). Therefore we get

$$A = \frac{3}{4 \varepsilon m},$$

so that from (2.15) we have

$$C = n \left(\frac{3}{4 \pi \varepsilon m}\right)^{3/2}.$$

- We obtained the Maxwell distribution

$$\varrho_0(p) = n \left(\frac{3}{4\pi\varepsilon m} \right)^{3/2} e^{-3\|p\|^2/(4m\varepsilon)}.$$

- Now assume that $\mathscr{U}$ is not zero. It is easy to verify that the Ansatz $\varrho(q,p) := \varrho_0(p)\,\sigma(q)$, for some function σ, still provides the vanishing of the r.h.s. of (2.9). From (2.5) we can write

$$\frac{d\varrho}{dt} = \frac{\varrho_0(p)}{m} \Big\langle \operatorname{grad}_q \sigma(q), p \Big\rangle - \sigma(q) \Big\langle \operatorname{grad}_p \varrho_0(p), \operatorname{grad}_q \mathscr{U}(q) \Big\rangle = 0, \quad (2.16)$$

which is a partial differential equation for the unknown function σ.

- Using the form of ϱ_0, equation (2.16) can be written as

$$\operatorname{grad}_q \sigma(q) + \frac{3}{2\varepsilon} \sigma(q) \operatorname{grad}_q \mathscr{U}(q) = 0,$$

whose solution is easily found:

$$\sigma(q) = D\, e^{-3\,\mathscr{U}(q)/(2\varepsilon)}.$$

Here the constant $D > 0$ is fixed by the normalization

$$\int_{\mathbb{R}^3} \varrho_0(p)\,dp \int_\Lambda \sigma(q)\,dq = N.$$

One finds

$$D = \left(\frac{1}{V} \int_\Lambda e^{-3\,\mathscr{U}(q)/(2\varepsilon)}\,dq \right)^{-1}.$$

The Theorem is proved. ∎

▶ Remarks:

- The equilibrium distributions (2.12) and (2.13) do not depend on the transition kernel τ.

- Once the transition kernel τ is defined, the mechanics of the collision does not depend on the identification of the particles. Indeed, the indices of the outgoing particles are assigned for convenience, but the symmetry properties of τ allow them to be interchanged, so that the outgoing particles are not only identical, but also *indistinguishable*.

2.3 Thermodynamics of a free ideal gas

▶ We consider an ideal gas without external force fields, so that the equilibrium probability distribution function is the Maxwell distribution (2.13):

$$\varrho_0(p) := n \left(\frac{3}{4\pi\varepsilon m} \right)^{3/2} e^{-3\|p\|^2/(4m\varepsilon)},$$

where

$$\varepsilon := \frac{1}{2m} \left\langle \|p\|^2 \right\rangle_{\varrho_0}. \tag{2.17}$$

2.3.1 Derivation of thermodynamic properties

▶ We show how to define and derive the thermodynamics of the system. In particular we show the validity of the (macroscopic) *free ideal gas law*

$$PV = N\kappa T.$$

- Let $d\Lambda$ be an infinitesimal area of the boundary $\partial\Lambda$. Particles exert on $d\Lambda$ an infinitesimal force which is by definition

$$dF := P\,d\Lambda. \tag{2.18}$$

- Every particle colliding with $d\Lambda$ is subject to a variation of its momentum in the direction normal to $d\Lambda$ and equal to twice the norm of the normal component p_n of its momentum preceding the collision.

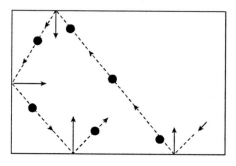

Fig. 2.3. Elastic collision of a particle against $\partial\Lambda$.

Newton's law implies that the infinitesimal force per particle exerted on $d\Lambda$ is obtained by multiplying $2\|p_n\|$ by the number of collisions experienced per unit time due to particles with momentum in the cell dp. This number is $(\|p_n\| \varrho_0(p)\, d\Lambda\, dp)/m$ (*collision frequency*). We finally need to integrate on the space of momenta which produce collisions, i.e., $\|p_n\| > 0$. Hence we find that the infinitesimal force exerted on $d\Lambda$ is

$$dF = \frac{2}{m}\, d\Lambda \int_{\|p_n\|>0} \|p_n\|^2\, \varrho_0(p)\, dp. \qquad (2.19)$$

- Comparing (2.18) and (2.19) we find

$$\begin{aligned} P &= \frac{2}{m} \int_{\|p_n\|>0} \|p_n\|^2 \varrho_0(p)\, dp = \frac{1}{m} \int_{\mathbb{R}^3} \|p_n\|^2 \varrho_0(p)\, dp \\ &= \frac{n}{m} \Big\langle \|p_n\|^2 \Big\rangle_{\varrho_0} . \end{aligned}$$

- The isotropy symmetry of the problem suggests that the particles move in random directions and, as a consequence, there is an equal probability of a particle moving in any direction. Therefore,

$$\Big\langle \|p_n\|^2 \Big\rangle_{\varrho_0} = \frac{1}{3} \Big\langle \|p\|^2 \Big\rangle_{\varrho_0} = \frac{2}{3}\, m\, \varepsilon,$$

where we used (2.17). Therefore the *pressure* of the gas can be expressed in terms of the average kinetic energy ε:

$$P = \frac{2}{3}\, n\, \varepsilon. \qquad (2.20)$$

- The *temperature* of the gas is defined as a quantity which is proportional to the average kinetic energy ε:

$$T := \frac{2\,\varepsilon}{3\,\kappa}.$$

It is useful to introduce the so called *inverse temperature*:

$$\beta :- \frac{1}{\kappa\, T},$$

so that we also write

$$\varepsilon = \frac{3}{2\,\beta}. \qquad (2.21)$$

- We can define the total *internal energy* of the gas as

$$E := N\varepsilon = \frac{3}{2}\, N\kappa\, T. \qquad (2.22)$$

Note that E is an extensive variable: it grows linearly with N.

- Comparing (2.20) and (2.22) there follows the free ideal gas law

$$PV = N\kappa T.$$

▶ Remarks:

- The introduction of the temperature T (or the inverse temperature β) allows us to write down the Maxwell distribution (2.13) in a suggestive way. Using (2.21) we can eliminate ε from (2.13) to get

$$\varrho_0(p) = n \left(\frac{\beta}{2\pi m} \right)^{3/2} e^{-\beta\, \mathfrak{h}(p)}, \tag{2.23}$$

where

$$\mathfrak{h}(p) := \frac{\|p\|^2}{2m}$$

is the canonical Hamiltonian of a free particle with mass m. It is not hard to prove that for the case of an ideal gas with external conservative forces, the Boltzmann-Maxwell distribution (2.12) can be written as

$$\varrho(q, p) = \alpha\, e^{-\beta\, \mathfrak{h}(q,p)},$$

where $\alpha > 0$ is a normalization factor and

$$\mathfrak{h}(q, p) := \frac{\|p\|^2}{2m} + \mathscr{U}(q).$$

Note that the collision forces do not contribute to $\mathfrak{h}(q, p)$, confirming the fact that in our assumptions these do not change the structure of the equilibrium, although they play a determining role in leading the system towards it.

- The definition of the total internal energy E allows us to complete the logical path from the microscopic model to the thermodynamics of the system. According to formulas (1.3) and (1.4) we can write (N is constant)

$$- P\, dV + dQ = dE, \tag{2.24}$$

where dQ is the quantity of heat exchanged with the exterior. This leads to the *first law of thermodynamics*.

2.3.2 *Entropy and convergence to thermodynamic equilibrium*

▶ Since there are no external fields we can assume that the non-equilibrium distribution function is a function depending on momentum p and time t, say $\varrho(p,t)$. We know that when the gas is in thermodynamic equilibrium the distribution is the Maxwell distribution (2.13). We now want to prove that condition (2.14), from which we derived (2.13), is not only sufficient but also necessary.

- Rewrite the Boltzmann transport equation (2.9) for $\varrho_1 := \varrho(p_1, t)$ as

$$\frac{d\varrho_1}{dt} = \frac{\partial \varrho_1}{\partial t} = \int_{\mathbb{R}^3} dp_2 \int_{\Sigma_{p_{\text{tot}}, E_{\text{tot}}}} \tau\,(p_1, p_2, \tilde{p}_1, \tilde{p}_2)\,\delta\,(p_1, p_2, \tilde{p}_1, \tilde{p}_2, t)\,d\tilde{p}_1\,d\tilde{p}_2, \quad (2.25)$$

where

$$\delta\,(p_1, p_2, \tilde{p}_1, \tilde{p}_2, t) := \varrho(\tilde{p}_1, t)\,\varrho(\tilde{p}_2, t) - \varrho(p_1, t)\,\varrho(p_2, t). \quad (2.26)$$

and

$$\Sigma_{p_{\text{tot}}, E_{\text{tot}}} := \left\{ (\tilde{p}_1, \tilde{p}_2) \in \mathbb{R}^6 \,:\, \tilde{p}_1 + \tilde{p}_2 = p_{\text{tot}},\, \frac{1}{2\,m}\left(\|\tilde{p}_1\|^2 + \|\tilde{p}_2\|^2 \right) = E_{\text{tot}} \right\},$$

with p_{tot} fixed by $p_1 + p_2$ and E_{tot} fixed by $\left(\|p_1\|^2 + \|p_2\|^2 \right) / (2\,m)$.

- Define the *Boltzmann functional* by

$$H(t) := \int_{\mathbb{R}^3} \varrho(p, t) \log \varrho(p, t)\,dp, \quad (2.27)$$

where the integral is assumed to be convergent. Recall that ϱ / N is a probability density. Indeed, formula (2.27) can be compared with the entropy function (1.7).

- At thermodynamic equilibrium we compute (see Lemma 2.1)

$$\begin{aligned}
H &:= \int_{\mathbb{R}^3} \varrho_0(p) \log \varrho_0(p)\,dp \\
&= 4\,\pi n \left(\frac{\beta}{2\,\pi\,m} \right)^{3/2} \log \left(n \left(\frac{\beta}{2\,\pi\,m} \right)^{3/2} \right) \int_0^{+\infty} \|p\|^2 e^{-\beta \|p\|^2 / (2\,m)}\,d\|p\| \\
&\quad - \frac{4\,\pi n\,\beta}{2\,m} \left(\frac{\beta}{2\,\pi\,m} \right)^{3/2} \int_0^{+\infty} \|p\|^4 e^{-\beta \|p\|^2 / (2\,m)}\,d\|p\|, \\
&= n \log \left(n \left(\frac{\beta}{2\,\pi\,m} \right)^{3/2} \right) - \frac{3}{2}\,n. \quad (2.28)
\end{aligned}$$

- Define the *entropy* of the gas (in a non-equilibrium state) by

$$S(t) := -\kappa\,V\,H(t) = -\kappa\,V \int_{\mathbb{R}^3} \varrho(p, t) \log \varrho(p, t)\,dp, \quad (2.29)$$

which is a function defined up to an additive constant. Note that the entropy is an extensive quantity: it grows with the volume when the average density n is fixed.

- At thermodynamic equilibrium we compute (see (2.28))

$$
\begin{aligned}
S := -\kappa V H &= -\kappa N \log\left(n \left(\frac{\beta}{2\pi m}\right)^{3/2}\right) + \frac{3}{2}\kappa N \\
&= \kappa N \log\left(\frac{V}{N}\left(\frac{4\pi m E}{3N}\right)^{3/2}\right) + \frac{3}{2}\kappa N, \qquad (2.30)
\end{aligned}
$$

where we used (2.22). Regarding S as a function of E and V and using (2.24), where $dQ = T\,dS$ (see (1.2)), we have (N is constant)

$$
dS = \frac{\partial S}{\partial E}\,dE + \frac{\partial S}{\partial V}\,dV = \frac{1}{T}\,dE + \frac{P}{T}\,dV.
$$

This allows us to recover important thermodynamics properties. Indeed we have

$$
\frac{\partial S}{\partial E} = \frac{3\kappa N}{2E} = \frac{1}{T}, \qquad \frac{\partial S}{\partial V} = \frac{N\kappa}{V} = \frac{P}{T}.
$$

Note that S given by (2.30) is an extensive variable.

▶ The next Theorem shows that the Boltzmann functional (2.27) is non-increasing along solution trajectories of the Boltzmann equation (2.25). The functional H is minimal at equilibrium. Let us mention that the claim which follows implies the *second law of thermodynamics* and it is the cornerstone of the *variational formulation of thermodynamics*.

Theorem 2.2 (Boltzmann H-Theorem)

If $\varrho(p,t)$ solves (2.25) then

$$
\frac{dH}{dt} \leq 0,
$$

where equality holds if and only if $\delta\left(p_1, p_2, \tilde{p}_1, \tilde{p}_2, t\right) = 0$ for all $t \geq 0$.

Proof. For notational convenience we omit the arguments $p_1, p_2, \tilde{p}_1, \tilde{p}_2, t$ of the transition kernel τ and of the function δ defined in (2.26), but we recall that τ is symmetric both w.r.t. $(p_1, p_2) \leftrightarrow (\tilde{p}_1, \tilde{p}_2)$ and $p_1 \leftrightarrow p_2$ and $\tilde{p}_1 \leftrightarrow \tilde{p}_2$, while δ is skew-symmetric w.r.t. $(p_1, p_2) \leftrightarrow (\tilde{p}_1, \tilde{p}_2)$ and symmetric w.r.t. $p_1 \leftrightarrow p_2$ and $\tilde{p}_1 \leftrightarrow \tilde{p}_2$.

- Fix $p = p_1$. Using (2.25) and (2.27) we get

$$
\begin{aligned}
\frac{dH}{dt} &= \int_{\mathbb{R}^3} \frac{\partial \varrho_1}{\partial t}\left(1 + \log \varrho(p_1, t)\right) dp_1 \\
&= \int_{\Omega} \tau\,\delta\left(1 + \log \varrho(p_1, t)\right) dp_1\, dp_2\, d\tilde{p}_1\, d\tilde{p}_2, \qquad (2.31)
\end{aligned}
$$

where Ω is the regular submanifold defined by those momenta satisfying (2.6) and (2.7).

- Similarly, for $p = p_2$, we get

$$
\begin{aligned}
\frac{dH}{dt} &= \int_{\mathbb{R}^3} \frac{\partial \varrho_2}{\partial t} \left(1 + \log \varrho(p_2, t)\right) dp_2 \\
&= \int_{\Omega} \tau \, \delta \left(1 + \log \varrho(p_2, t)\right) dp_1 \, dp_2 \, d\tilde{p}_1 \, d\tilde{p}_2.
\end{aligned} \tag{2.32}
$$

- The sum of (2.31) and (2.32) gives

$$
2 \frac{dH}{dt} = \int_{\Omega} \tau \, \delta \left(2 + \log \left(\varrho(p_1, t) \, \varrho(p_2, t)\right)\right) dp_1 \, dp_2 \, d\tilde{p}_1 \, d\tilde{p}_2.
$$

- The symmetry of τ and the skew-symmetry of δ under $(p_1, p_2) \leftrightarrow (\tilde{p}_1, \tilde{p}_2)$ imply that also

$$
2 \frac{dH}{dt} = - \int_{\Omega} \tau \, \delta \left(2 + \log \left(\varrho(\tilde{p}_1, t) \, \varrho(\tilde{p}_2, t)\right)\right) dp_1 \, dp_2 \, d\tilde{p}_1 \, d\tilde{p}_2
$$

holds true.

- The sum of the last two equations gives

$$
4 \frac{dH}{dt} = \int_{\Omega} \tau \, \delta \, w \left(p_1, p_2, \tilde{p}_1, \tilde{p}_2, t\right) dp_1 \, dp_2 \, d\tilde{p}_1 \, d\tilde{p}_2. \tag{2.33}
$$

where

$$
w \left(p_1, p_2, \tilde{p}_1, \tilde{p}_2, t\right) := \log(\varrho(p_1, t) \, \varrho(p_2, t)) - \log(\varrho(\tilde{p}_1, t) \, \varrho(\tilde{p}_2, t)).
$$

- The r.h.s. of (2.33) is non-positive, since for each pair of positive numbers (x, y) we have $(y - x)(\log x - \log y) \leqslant 0$, with equality if and only if $x = y$.

The Theorem is proved. ∎

▶ Theorem 2.2 has three relevant consequences.

Corollary 2.1

1. *Condition* $\delta \left(p_1, p_2, \tilde{p}_1, \tilde{p}_2, t\right) = 0$ *for all* $t \geqslant 0$ *is a necessary and sufficient condition for the distribution* $\varrho(p, t)$ *to be an equilibrium distribution function.*

2. *A free ideal gas in a non-equilibrium state corresponding to a distribution function* $\varrho(p, 0)$ *at* $t = 0$ *converges asymptotically for* $t \to +\infty$ *towards an equilibrium state corresponding to the Maxwell distribution* (2.13).

3. *The entropy function* $S(t)$ *grows until equilibrium is achieved (second law*

> **of thermodynamics**). *At equilibrium the entropy is maximal.*

Proof. We give a proof for all claims.

1. Stationarity of equilibrium distribution functions implies $dH/dt = 0$, from which $\delta\left(p_1, p_2, \widetilde{p}_1, \widetilde{p}_2, t\right) = 0$ for all $t \geqslant 0$ necessarily follows. Sufficiency is obvious.

2. It follows from the monotonicity of H.

3. It follows from the monotonicity of H and from formula (2.29) defining the entropy.

All claims are proved.   ∎

▶ There are several paradoxes stemming from the interpretation of Theorem 2.2 (1871) and its consequences. The most evident paradox is the manifestation of the (time) *non-reversibility* and *non-recurrency* of the process achieving macroscopic equilibrium, opposed to the (time) reversible and recurrent behavior of mechanics governing the microscopic dynamics of the system.

- The scenario is indeed much more general and, from the point of view of constructing mathematical models for any kind of observable process, much more fundamental. All theories of microscopic physics are governed by laws that are invariant under reversal of time: the evolution of the system can be traced back into the past by the same evolution equation that governs the prediction into the future. Processes on macroscopic scales, on the other hand, are manifestly irreversible.

- In particular, classical thermodynamics provides accurate quantitative descriptions of observables and irreversible phenomena without reference to the underlying microphysics. This raises at least two questions.

    1. Since we have now two presumably accurate, quantitative mathematical models of the same system, how does one embed into the other?
    2. If such a relation can be established, how can it even be that one of these descriptions is reversible whereas the other is not?

The "naive" answer in the context of kinetic theory of gases, is that the macroscopic laws are a genuine statistical description: they represent the most probable behavior of the system. At the same time, most microscopic realizations of a macroscopic state remain close to the most probable behavior for a long, but finite interval of time. The rigorous answer, which provides quantitative definiteness and mathematical rigor, is, in most cases, truly long and difficult.

▶ As a matter of fact, Boltzmann's work faced severe criticism, essentially on two levels.

1. *Loschmidt's criticism* (1871). Soon after Boltzmann published his Theorem 2.2, J.J. Loschmidt objected that it should not be possible to deduce an irreversible process from deterministic and time reversible mechanics. The origin of this contradiction lies the Stosszahlansatz, i.e., it is acceptable for all the particles to be considered independent and uncorrelated. In some sense, this controversial Ansatz breaks time reversal symmetry and therefore begs the question. Once the particles are allowed to collide, their velocity directions and positions in fact do become correlated (however, these correlations are encoded in an extremely complex manner). Boltzmann's reply to Loschmidt was to admit the possible occurence of these states, but noting that these sorts of states were so rare as to be impossible in practice. Boltzmann would go on to sharpen this notion of the "rarity" of states, resulting in his famous equation, his entropy formula (1877).

2. *Zermelo's criticism* (1896). E.F. Zermelo gave a new proof of "Poincaré recurrence Theorem" and he proved its applicability to the situation considered by Boltzmann. "Poincaré recurrence Theorem" states that a flow defined on a compact phase space which preserves the volume of the phase space must eventually return to its initial state within arbitrary precision for almost all initial data. Zermelo noted a further problem with Theorem 2.2: if the functional H is at any time not a minimum, then by "Poincaré recurrence Theorem", the non-minimal H must recur (though after some extremely long time). Boltzmann admitted that these recurring rises in H technically would occur, but pointed out that, over long times, the system spends only a tiny fraction of its time in one of these recurring states. Since H is a mechanically defined variable that is not conserved, then like any other such variable (pressure, etc.) it will show *thermal fluctuations*. This means that H regularly shows spontaneous increases from the minimum value. Technically this is not an exception to Theorem 2.2, since the claim was only intended to apply for a gas with a very large number of particles. These fluctuations are only perceptible when the system is small. If H is interpreted as entropy, as Boltzmann intended, then this can be seen as a manifestation of fluctuations.

Example 2.1 (*Maxwell Gedankenexperiment*)

Suppose to have a box partitioned into two chambers, say A and B. Suppose that A contains a gas, while B is empty. We make a hole in the wall separating A and B. Then it is reasonable to expect that after some time the gas will be uniformly distributed in A and B.

- We are under the conditions of "Poincaré recurrence Theorem". Indeed, the volume of the box is finite and, assuming elastic and regular interactions between particles, the conservation of the global energy assures that particle velocities are bounded. Then the phase space is a compact space and "Poincaré recurrence Theorem" says that there exists a time T for which all particles constituting the gas will come back to a configuration which is close to the initial configuration, that is all molecules in the chamber A!

- The resolution of this paradoxical situation lies in the fact that T is longer than the duration of the solar system's existence (...billions of years). Furthermore, one of the assumption we used in this ideal experiment is that the system is isolated (i.e., no external perturbations are admitted). This assumption is not realistic, especially on long times.

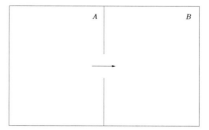

Fig. 2.4. Maxwell Gedankenexperiment.

2.4   The Kac ring model

▶ The *Kac ring* is a simple and explicitly solvable model which illustrates the transition from a microscopic and time-reversible picture to a macroscopic and time-irreversible one. The presentation which follows is taken from the paper "Boltzmann's Dilemma: An Introduction to Statistical Mechanics via the Kac Ring" by G.A. Gottwald, M. Oliver, SIAM Review, 51/3, 2009.

▶ The model is constructed as follows.

- N sites are arranged around a circle, forming a one-dimensional periodic lattice. Neighboring sites are joined by an edge. Each site is occupied by either a black ball or a white ball. Moreover, $n < N$ edges carry a marker.

- The system evolves on a discrete set of clock-ticks $t \in \mathbb{Z}$ from state t to state $t + 1$ according to the following rule: each ball moves clockwise to the neighboring site. When a ball passes a marker, its color changes.

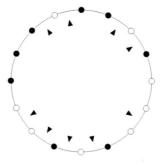

Fig. 2.5. A Kac ring ([GoOl]).

▶ Note that the microscopic dynamics of the Kac ring is:

- time-reversible: when the direction of movement along the ring is reversed, the balls retrace their past color sequence.

- recurrent: after N clock ticks, each ball has reached its initial site and changed color n times. Thus, if n is even, the initial state recurs; if n is odd, it takes at most $2N$ clock ticks for the initial state to recur.

▶ We now describe the macroscopic dynamics of the model:

- Let $B = B(t)$ be the total number of black balls and $b = b(t)$ the number of black balls just in front of a marker. Let $W = W(t)$ be the number of white balls and $w = w(t)$ the number of white balls in front of a marker. Then:

$$B(t+1) = B(t) + w(t) - b(t), \qquad W(t+1) = W(t) + b(t) - w(t). \qquad (2.34)$$

- Define $\delta = \delta(t) := B(t) - W(t)$. Then

$$\delta(t+1) = B(t+1) - W(t+1) = \delta(t) + 2(w(t) - b(t)). \qquad (2.35)$$

- Note that B, W, δ are macroscopic quantities, describing a global feature of the system while b, w contain local information about individual sites: they cannot be computed without knowing the location of each marker and the color of the ball at every site. A key feature is that the evolution of the global quantities B, W, δ is not computable only from macroscopic state information. More concretely, it is not possible to eliminate b and w from (2.34) (closure problem). To solve this problem and thus obtain a macroscopic description of the model we need an additional assumption, analogous to the Stosszahlansatz.

- When the markers are distributed at random, the probability that a particular site is occupied by a marker is given by

$$\mathsf{P} := \frac{n}{N} = \frac{b}{B} = \frac{w}{W} \in (0,1). \qquad (2.36)$$

For an actual realization of the Kac ring these relations will generally not be satisfied. However, by assuming that they hold anyway, we can overcome the closure problem. This assumption is the analogue of the Stosszahlansatz. It effectively disregards the history of the system evolution: there is no memory of where the balls originated and which markers they passed up to time t. We assume that this hypothesis represents the typical behavior of large sized rings.

- Using (2.36) we can derive our macroscopic description. Indeed, equation (2.35) now takes the form

$$\delta(t+1) = \delta(t) + 2\,\mathsf{P}\,(W(t) - B(t)) = (1 - 2\,\mathsf{P})\,\delta(t),$$

which is solved by

$$\delta(t) = (1 - 2\,\mathsf{P})^t \delta(0). \qquad (2.37)$$

▶ Remarks:

- Equation (2.37) plays the role of Boltzmann transport equation. It cannot describe the dynamics of one particular ring exactly. For instance, δ is generically not an integer anymore.

- Since $0 < P < 1$, we see that $\delta \to 0$ as $t \to +\infty$. Contrary to what we know about the microscopic dynamics, the magnitude of δ in (2.37) is monotonically decreasing and therefore time irreversible (*Loschmidt's criticism*).

- The initial state cannot recur, again in contrast to the microscopic dynamics which has a recurrence time of at most $2N$ (*Zermelo's criticism*).

▶ Our task is to give a meaning to the macroscopic evolution equation (2.37) on the basis of the microscopic dynamics. Boltzmann's point of view would be that the macroscopic law (2.37) can only be valid in a statistical sense, referring to the most probable behavior of a member of a large *ensemble of systems* rather than to the exact behavior of any member of the ensemble. Therefore the microscopic dynamics of our model admits a natural probabilistic interpretation.

- By an ensemble of Kac rings we mean a collection of rings with the same number of sites.

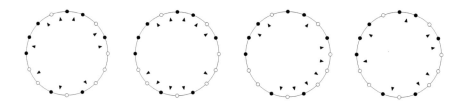

Fig. 2.6. Ensembre of Kac rings ([GoOl]).

Each member of the ensemble has the same initial configuration of black and white balls. The markers, however, are placed at random in such a way that the probability that an edge is occupied by a marker is P.

- Let f denote some function of the configuration of markers and f_j denote the value of f for the j-th member of the ensemble.

- The *ensemble average* $\langle f \rangle$ is defined as the arithmetic mean of f over a large number of realizations:

$$\langle f \rangle := \lim_{M \to +\infty} \frac{1}{M} \sum_{j=1}^{M} f_j. \tag{2.38}$$

- In the language of probability theory, each particular configuration of markers is referred to as outcome from among the sample space X of all possible configurations of markers. The process of choosing a random configuration of markers is a trial. It is always assumed that trials are independent. In this framework $f : X \to \mathbb{R}$ is a *random variable*.

- Thus $\langle f \rangle$ is nothing but the expected value of a random variable, and can be computed as follows. As the system is finite, f will take one of $x_1, \dots, x_\ell$ possible values ("macrostates") with corresponding probabilities $p_1, \dots, p_\ell$. Then:

$$\langle f \rangle = \sum_{j=1}^{\ell} p_j \, x_j. \tag{2.39}$$

The identification of (2.38) and (2.39) is due to the definition of probability p_j of the event $f = x_j$ as its relative frequency of occurrence in a large number of trials.

▶ According to the previous observations we give a formulation of the microscopic dynamics which can justify the macroscopic law (2.37).

- Let $\chi_i = \chi_i(t)$ denote the color of the ball occupying the i-th lattice site at time t, with $\chi_i = +1 \equiv black$ and $\chi_i = -1 \equiv white$. Further, let $m_i = -1$ denote the presence and $m_i = +1$ denote the absence of a marker on the edge connecting sites i and $i + 1$.

- Then the recurrence relation for stepping from t to $t + 1$ reads

$$\chi_{i+1}(t+1) = m_i \, \chi_i(t),$$

which makes sense for any $i, t \in \mathbb{Z}$, if we impose periodic boundary conditions $\chi_1 \equiv \chi_{N+1}$, $m_1 \equiv m_{N+1}$. Then

$$\delta(t) = \sum_{i=1}^{N} \chi_i(t) = \sum_{i=1}^{N} m_{i-1} \chi_{i-1}(t-1) = \sum_{i=1}^{N} m_{i-1} m_{i-2} \chi_{i-2}(t-2)$$

$$= \cdots = \sum_{i=1}^{N} m_{i\,1} m_{i\,2} \cdots m_{i-t} \chi_{i-t}(0).$$

Note that $\delta(2N) = \delta(0)$.

- We wish to compute the evolution of the ensemble average $\langle \delta(t) \rangle$. Since averaging involves taking sums and since only the marker positions, but not the initial configuration of balls, may differ across the ensemble, we can extract the sum and the $\chi_i(0)$ from the average and obtain

$$\langle \delta(t) \rangle = \sum_{i=1}^{N} \langle m_{i-1} m_{i-2} \cdots m_{i-t} \rangle \chi_{i-t}(0). \tag{2.40}$$

- Since all lattice edges have equal probability of carrying a marker, the average (2.40) must be invariant under index shifts. Hence $\langle m_{i-1}m_{i-2}\dots m_{i-t}\rangle = \langle m_1 m_2 \dots m_t\rangle$ so that

$$\langle \delta(t)\rangle = \langle m_1 m_2 \dots m_t\rangle \sum_{i=1}^{N} \chi_{i-t}(0) = \langle m_1 m_2 \dots m_t\rangle \delta(0). \qquad (2.41)$$

- Our task is to find an explicit expression for $\langle m_1 m_2 \dots m_t\rangle$, a quantity which only depends on the distribution of the markers, but not on the balls. We distinguish two cases:

  1. $0 \leqslant t < N$. There are no periodicities: all factors $m_1, \dots, m_t$ are independent. The value of the product is 1 for an even number of markers, and -1 for an odd number of markers. Thus, (2.39) takes the form

$$\langle m_1 m_2 \dots m_t\rangle = \sum_{j=0}^{t} (-1)^j p_j(t),$$

where $p_j(t)$ denotes the probability of finding j markers on t consecutive edges. The markers follow a *binomial distribution*, so that

$$p_j(t) := \binom{t}{j} \mathsf{P}^j (1 - \mathsf{P}^{t-j}).$$

From the "Binomial Theorem" we get

$$\langle m_1 m_2 \dots m_t\rangle = \sum_{j=0}^{t} \binom{t}{j} (-\mathsf{P})^j (1 - \mathsf{P}^{t-j}) = (1 - 2\,\mathsf{P})^t,$$

so that from (2.41) we have

$$\langle \delta(t)\rangle = (1 - 2\,\mathsf{P})^t \delta(0),$$

the same expression (2.37) we got through our initial Stosszahlansatz. This computation shows that the relatively crude Stosszahlansatz may be related to the average over a statistical ensemble. In general, however, one cannot expect exact identity. Indeed, even for the Kac ring, the next case shows that when $t > N$, the two concepts diverge.

  2. $N \leqslant t < 2N$. Now the balls may pass some markers twice, and we have to explicitly account for these periodicities:

$$\langle m_1 m_2 \dots m_t\rangle = \langle m_{t+1}m_{t+2}\dots m_{2N}\rangle = \langle m_1 m_2 \dots m_{2N-t}\rangle.$$

The first equality is a consequence of the N-periodicity of the lattice, which implies that $m_1 m_2 \dots m_{2N} = 1$. The second equality is due again to the invariance of the average under an index shift. Now a simple computation gives

$$\langle \delta(t)\rangle = (1 - 2\,\mathsf{P})^{2N-t}\delta(0).$$

As the exponent on the right hand side is negative on the interval $N \leqslant t < 2N$, the ensemble average $\langle \delta(t) \rangle$ increases on this interval and, in particular, returns to its initial value for $t = 2N$.

▶ We have shown that *the Stosszahlansatz leads to the macroscopic equation (2.37) that represents the averaged behavior of an ensemble of Kac rings for times* $0 \leqslant t < N$. A more rigorous analysis (based on the variance of δ) shows that for short times and large N the average behavior is indeed *typical*.

▶ We now give a characterization of the entropy of the Kac ring. We start with the following two important features of our system:

1. The system is made up of N independent identical components which can be in one of two possible states.

2. The macroscopic observable is proportional to the number of components in each state.

▶ Then the entropy is computed according to the following procedure.

• We introduce the *partition function* $\mathcal{Z}$ of the system, defined as the number of microstates for a given macrostate. The macrostate is fully specified by δ or, equivalently, by the number of black balls $B = (N + \delta)/2$ or the number of white balls $W = (N - \delta)/2$.

• Then the state with B black balls and $W = N - B$ white balls can be realized in

$$\mathcal{Z}(B) := \frac{N!}{B!W!}$$

different ways.

• The logarithm of $\mathcal{Z}$, turns out to be more useful because it scales approximately linearly with system size when N is large, as we shall show below. This motivates the definition of the *entropy*

$$S := \log \mathcal{Z}.$$

The entropy S is a function of the macrostate δ only.

• Let $p := B/N$ denote the probability that a site carries a black ball and $q := 1 - p = W/N$ the probability that a site carries a white ball. Then, by using Stirling approximation (2.11) we get, for large N,

$$S = \log \left(\frac{N!}{(pN)!(qN!)} \right) \approx -N(p \log p + q \log q),$$

which is an extensive quantity given by the product of two terms, the first of which depends only on the size of the system, the second only on the macroscopic state.

▶ Final remarks:

- We considered a system with a very large microscopic state space that can be decomposed into a large number of simple interacting subsystems: the balls and markers (or the particles in a gas in Boltzmann's description). The dynamical description at this level is deterministic and time reversible.

- The microscopic state is assumed to be non-observable. We thus introduce a "coarse-graining function", a many-to-one map from the microscopic state space into a much smaller macroscopic state space. The output of the "coarse-graining function" is the experimentally accessible quantity δ. The main problem is to understand the time evolution of δ.

- Given that the "coarse-graining function" is highly non-invertible, we resort to a statistical description. For a given initial $\delta(0)$, we construct an ensemble of systems such that the corresponding macrostate, the expected value of the "coarse-graining function" applied to the members of the ensemble, matches $\delta(0)$. In the absence of further information, we must assume that all constituent subsystems are statistically independent. We can then describe the evolution of δ in two different ways.

  1. *Newton's approach.* The dynamics is given by the full evolution at microscopic level. We evolve each member of the ensemble of microstates up to some final time, apply the "coarse-graining function" to each member of the ensemble, and finally compute the statistical moments of the resulting distribution of macrostates. Macroscopic information is extracted a posteriori by computing relevant averages. Then a single performance of the experiment will, with high probability, evolve close to the ensemble mean. The Hamiltonian formalism of classical Newtonian mechanics provides, in principle, a correct description, but it may be theoretically and computationally intractable.

  2. *Boltzmann's approach.* A crude Stosszahlansatz simplifies the model and evolves a macrostate. It is based on the fact that under the assumption of statistical independence of the subsystems, it is usually easy to predict the macroscopic mean after one time step. However, the interaction of subsystems during a first time step will generally destroy their statistical independence. Still, we might pretend that, after each time step, all subsystems are still statistically independent. In general, the resulting macroscopic dynamics will differ from the predictions of Newton's dynamics after more than one time step. For the Kac ring, they differ when $t > N$.

- Both approaches break the time-reversal symmetry of the microscopic dynamics, as the "coarse-graining function" from an ensemble of microstates to the macroscopic ensemble mean is invertible if and only if the constituent subsystems are statistically independent. Hence, the loss of statistical independence defines a macroscopic *arrow of time*. This argument solves Loschmidt's paradox.

- The Stosszahlansatz in the Boltzmann approach is an approximation which depends on weak statistical dependence of the subsystems. However, as time passes, interactions will increase statistical dependence. Thus, we cannot expect that the approximate macroscopic mean remains a faithful representation of the recurrent microscopic dynamics over a long period of time. The validity of "Poincaré recurrence Theorem" for dynamical systems does not imply its applicability to ensembles. This argument solves Zermelo's paradox.

2.5   Exercises

Ch2.E1 Consider a closed and isolated system of $N \gg 1$ distinguishable but independent identical particles. Each particle can exist in one of two states with energy difference $\varepsilon > 0$. Given that m particles are in the excited state, the total energy of the system is $m\,\varepsilon$ with a degeneracy of

$$\mathcal{Z}(N, m) := \frac{N!}{(N - m)!\, m!}.$$

(a) Give a combinatorial interpretation of $\mathcal{Z}(N, m)$.

(b) Define the entropy of the system by

$$S := \kappa \log \mathcal{Z}(N, m),$$

where κ is the Boltzmann constant. Determine S in the Stirling approximation. From now on work in this approximation.

(c) The absolute temperature of the system is defined by

$$T := \left(\frac{\partial S}{\partial E} \right)^{-1},$$

where E is the total energy. Compute the inverse temperature $\beta := (\kappa T)^{-1}$.

(d) Find the density of excited states m/N as a function of β.

(e) Use result (d) to write the entropy as a function of β.

(f) Determine the entropy in the limit $T \to 0$.

~~~~~~~~~~~~~~~~~~~~~~~~~~~~~~~~~~~

Ch2.E2 Consider a free ideal gas at equilibrium described by the Maxwell distribution:

$$\varrho_0(p) := n \left(\frac{\beta}{2\pi m} \right)^{3/2} e^{-\beta \|p\|^2 / (2m)}. \tag{2.42}$$

Here $\beta := (\kappa T)^{-1}$, m is the mass of the particles and $n := N/V$ is the number of particles per unit volume.

(a) Compute the average velocity of the particles, $\langle \|v\| \rangle_{\varrho_0} := \langle \|p\| \rangle_{\varrho_0} / m$.

Let $\delta > 0$ be the diameter of each particle. Consider pairs of colliding particles with momenta p_1 and p_2. Choosing a reference frame translating with one of the particle, the frequency of collisions per unit volume is defined by the positive number

$$\nu := \frac{\pi \delta^2}{m} \int_{\mathbb{R}^6} \|p_1 - p_2\|\, \varrho_0(p_1)\, \varrho_0(p_2)\, dp_1\, dp_2.$$

Since every collision involves only two particles, the total number of collisions to which a particle is subject per unit time can be found by dividing 2ν by the density n of particles.

(b) Compute explicitly ν.

(c) Prove that the mean-free path of each particle, defined by

$$\lambda := \frac{n}{2\nu} \langle \|v\| \rangle_{\varrho_0},$$

is given by

$$\lambda = \frac{1}{2\sqrt{2}} \frac{1}{\pi \delta^2 n}.$$

Give a rough estimate of λ for realistic values of δ and n. (Note that λ does not depend on the temperature!)

~~~~~~~~~~~~~~~~~~~~~~~~~~~~~~~

Ch2.E3 Consider an ideal gas in a box $[0, L]^3$ at equilibrium described by a Boltzmann-Maxwell distribution of the form

$$\varrho(q, p) := \varrho_0(p)\, \sigma(q),$$

where ϱ_0 is the Maxwell distribution (2.42) and σ is a function to be determined. Assume that the gas is subject to an external conservative force whose potential energy is

$$\mathscr{U}(q) := \mathscr{U}_0 \cos\left(\frac{2\pi \ell q_1}{L}\right), \qquad \mathscr{U}_0 > 0,\ \ell \in \mathbb{N}.$$

Here $q_1 \in [0, L]$ is the first component of the vector q.

(a) Find the function σ and determine an approximated formula for σ in the limit $\beta\,\mathscr{U}_0 \ll 1$.

(b) Prove that the total internal energy,

$$E := N \left\langle \frac{\|p\|^2}{2m} + \mathscr{U}(q) \right\rangle_\varrho,$$

is given by

$$E = \frac{3}{2} N \kappa T - N\, \mathscr{U}_0\, \frac{I_1(\beta\,\mathscr{U}_0)}{I_0(\beta\,\mathscr{U}_0)},$$

where the functions I_0 and I_1 are modified Bessel functions (see hint below). Note that E is the total internal energy of a free ideal gas if the external force is switched off.

(*Hint: The following integral representation and series expansion of the modified Bessel functions of the first kind are useful:*

$$I_n(z) = \frac{1}{\pi} \int_0^\pi e^{z\cos\theta} \cos(n\,\theta)\, d\theta = \left(\frac{z}{2}\right)^n \sum_{j\geqslant 0} \frac{(z^2/4)^j}{j!\,(n+j)!},$$

with $z \in \mathbb{C},\ n \in \mathbb{N}$)

~~~~~~~~~~~~~~~~~~~~~~~~~~~~~~~

Ch2.E4 Consider a box of volume V containing a free ideal gas. Momenta of particles are distributed according to the Maxwell distribution:

$$\varrho_0(p) := n \left(\frac{1}{2\pi m \kappa T_0}\right)^{3/2} e^{-\|p\|^2/(2m\kappa T_0)}.$$

Here T_0 is the temperature, m is the mass of the particles and $n_0 := N/V$ is the number of particles per unit volume. At time $t = 0$ a very small hole of area σ is made on one of the walls, allowing particles to escape into a surrounding vacuum. Suppose that the outward normal to the hole is the positive q_1 direction. For $t > 0$ we assume that the gas inside the box is always in equilibrium so that the distribution function is given by

$$\varrho_0(p, t) := n \left(\frac{1}{2\pi m \kappa T}\right)^{3/2} e^{-\|p\|^2/(2m\kappa T)},$$

where $n = n(t)$ and $T = T(t)$, with $T(0) = T_0$ and $n(0) = n_0$.

(a) The number of particles passing through the hole per unit time is given by

$$\dot{N}_{\text{out}} = \frac{\sigma}{m} \int_0^{+\infty} dp_1 \int_{-\infty}^{+\infty} dp_2 \int_{-\infty}^{+\infty} dp_3 \, p_1 \, \varrho_0(p,t).$$

Justify the above formula and prove that it implies the differential equation

$$\dot{n} = -A \, n \, T^{1/2}, \qquad A := \frac{\sigma}{V} \sqrt{\frac{\kappa}{2\pi m}}. \tag{2.43}$$

(*Hint: Note that* $\dot{n} = -\dot{N}_{\text{out}}$)

(b) The rate at which total internal energy is carried out through the hole by escaping particles is given by

$$\dot{E}_{\text{out}} = \frac{\sigma}{m} \int_0^{+\infty} dp_1 \int_{-\infty}^{+\infty} dp_2 \int_{-\infty}^{+\infty} dp_3 \, p_1 \left(\frac{\|p\|^2}{2m} \right) \varrho_0(p,t).$$

Justify the above formula and prove that it implies the differential equation

$$\dot{n} \, T + n \, \dot{T} = -\frac{4}{3} A \, n \, T^{3/2}. \tag{2.44}$$

(*Hint: Note that* $\dot{E} = -\dot{E}_{\text{out}}$, *where*

$$E(t) := \frac{3}{2} \kappa \, V \, n \, T$$

is the total internal energy of the gas at time t)

(c) Use (2.43) and (2.44) to find explicitly the temperature $T = T(t)$ and the number of particles per unit volume $n = n(t)$.

～～～～～～～～～～～～～～～～～～

Ch2.E5 Consider a one-dimensional lattice $\mathbb{Z}$ with lattice constant α. A particle transits from a site to a nearest-neighbor site every τ seconds. The probabilities of transiting to the left and to the right are p and $1 - p$ respectively. Find the average position of the particle at time $N\tau$, $N \gg 1$.

3

Gibbsian Formalism for Continuous Systems at Equilibrium

3.1 Introduction

▶ From "The value of science" (1905) by H. Poincaré: *"A drop of wine falls into a glass of water; whatever may be the law of the internal motion of the liquid, we shall soon see it colored of a uniform rosy tint, and however much from this moment one may shake it afterwards, the wine and the water do not seem capable of again separating. Here we have the type of the irreversible physical phenomenon: to hide a grain of barley in a heap of wheat, this is easy; afterwards to find it again and get it out, this is practically impossible. All this Maxwell and Boltzmann have explained; but the one who has seen it most clearly, in a book too little read because it is a little difficult to read, is Gibbs, in his Elementary Principles of Statistical Mechanics."*

▶ On the basis of Boltzmann's results, J.W. Gibbs (1839-1903) proposed a new approach for the study equilibrium states of systems with many degrees of freedom, with the aim of deducing their thermodynamic behavior starting from their Hamiltonian description.

▶ In our study we restrict our attention to a theoretical *gas of particles* at thermodynamic equilibrium, so defined:

1. *Hard spheres.* N (say $N \approx 6.02 \times 10^{23}$, *Avogadro number*) identical hard spheres (radius r, mass m) without internal structure contained in a bounded region $\Lambda \subset \mathbb{R}^3$, $\text{Vol}(\Lambda) = V$, at standard conditions of temperature T and pressure P. The volume V is not necessarily constant in time. Note that Λ can be seen as a real smooth submanifold of $\mathbb{R}^3$ which can be locally modeled over some Euclidean space with dimension $1 \leqslant \ell \leqslant 3$. For concrete models, such as the free ideal gas, we will consider $\ell = 3$.

2. *Hamiltonian formulation.* To simplify notation we work at a local level, thus adopting a canonical description of Hamiltonian mechanics.

   (a) We assign (time-dependent, $t \in \mathbb{R}$) generalized canonical coordinates $(q_i, p_i) \in \Lambda \times \mathbb{R}^\ell$, with $i = 1, \ldots, N$, to each particle. More precisely, the phase space of the particle is given by the 2ℓ-dimensional cotangent bundle of Λ, which is a symplectic manifold. As long as global coordinates do not play a relevant role we use a notation which is a bit sloppy (but standard) and we consider a single-particle phase space as a subset of $\mathbb{R}^{2\ell}$.

47

(b) The global $2\ell N$-dimensional canonical Hamiltonian *phase space*,

$$\Omega := \left(\Lambda \times \mathbb{R}^\ell \right)^N,$$

is parametrized by (time-dependent) coordinates

$$x := (q_1, \ldots, q_N, p_1, \ldots, p_N) \in \Omega,$$

where each q_i and p_i, $i = 1, \ldots, N$, has ℓ scalar components. We denote by

$$dx := dq_1 \cdots dq_N \, dp_1 \cdots dp_N$$

the infinitesimal volume element of Ω, i.e., the volume form of Ω. Therefore, if Ω_0 is a Lebesgue measurable subset of Ω we denote its Lebesgue measure (or volume), by

$$\mathrm{Vol}(\Omega_0) := \int_{\Omega_0} dx.$$

We also have

$$V^N = (\mathrm{Vol}(\Lambda))^N := \int_{\Lambda^N} dq_1 \cdots dq_N.$$

(c) We assume that the gas is governed by the following Hamiltonian:

$$\mathcal{H}(x) := \sum_{i=1}^{N} \frac{\|p_i\|^2}{2m} + \sum_{1 \leqslant i < j \leqslant N} \mathcal{U}(q_i - q_j) + \sum_{i=1}^{N} \mathcal{U}_{\mathrm{ext}}(q_i), \qquad (3.1)$$

where $\mathcal{U}$ is a *2-body interaction potential energy* and $\mathcal{U}_{\mathrm{ext}}$ is an *external potential energy* which describes the interaction between particles and some external (conservative) field. Our Hamiltonian system has ℓN *degrees of freedom*. This number can be reduced if there exist some additional integrals of motion, Poisson involutive with $\mathcal{H}$ and functionally independent on $\mathcal{H}$.

(d) The system is governed by the following system of $2\ell N$ canonical Hamilton equations:

$$\dot{x} = \mathbf{J} \, \mathrm{grad}_x \mathcal{H}(x),$$

where $\mathbf{J}$ is the canonical symplectic matrix

$$\mathbf{J} := \begin{pmatrix} \mathbf{0}_{\ell N} & \mathbf{1}_{\ell N} \\ -\mathbf{1}_{\ell N} & \mathbf{0}_{\ell N} \end{pmatrix}.$$

Some initial condition $x_0 := x(0) \in \Omega$ is prescribed.

(e) We assume completeness of the total Hamiltonian vector field, so that the total Hamiltonian flow $\Phi_t : \mathbb{R} \times \Omega \to \Omega$ is a one-parameter global Lie group of diffeomorphisms on Ω. The infinitesimal generator of Φ_t is

$$\mathbf{v} := \sum_{i=1}^{2\ell N} h_i(x) \frac{\partial}{\partial x_i}, \tag{3.2}$$

where $h_i(x)$ is the i-th component of

$$h(x) := \mathbf{J} \operatorname{grad}_x \mathcal{H}(x) = \frac{d}{dt}\bigg|_{t=0} \Phi_t(x).$$

We recall that Φ_t is energy-preserving, i.e., $\mathcal{H}$ is an integral of motion:

$$\mathbf{v}[\mathcal{H}(x)] = \frac{d}{dt}\bigg|_{t=0} (\mathcal{H} \circ \Phi_t)(x) = 0 \qquad \forall x \in \Omega,$$

or, equivalently,

$$(\mathcal{H} \circ \Phi_t)(x) = \mathcal{H}(\Phi_t(x)) = \mathcal{H}(x) \qquad \forall x \in \Omega, t \in \mathbb{R}.$$

Furthermore Φ_t, at every fixed t, is a symplectic transformation which preserves the canonical symplectic 2-form and the volume form dx, i.e., Φ_t preserves the Lebesgue measure of Ω.

Example 3.1 (*Gases of particles*)

1. Consider a system of N identical non-interacting particles with mass $m > 0$ contained in a cube of side $2L$ described by the Hamiltonian

$$\mathcal{H}(x) := \frac{1}{2} \sum_{i=1}^{N} \left(\frac{\|p_i\|^2}{m} + m\,\omega^2 \|q_i\|^2 \right), \qquad \omega \in \mathbb{R}.$$

Here the phase space of the system is $\Omega := ([-L, L]^3 \times \mathbb{R}^3)^N$. Note that the gravitational potential energy is neglected. This is indeed the Hamiltonian of a system of N three-dimensional independent harmonic oscillators. One could also consider a gas of N non-interacting particles (one-dimensional independent harmonic oscillators) distributed on a line of length L with Hamiltonian

$$\mathcal{H}(x) := \frac{1}{2} \sum_{i=1}^{N} \left(\frac{p_i^2}{m} + m\,\omega^2 q_i^2 \right), \qquad \omega \in \mathbb{R}.$$

Here the phase space of the system is $\Omega := ([0, L] \times \mathbb{R})^N$.

2. Consider a gas of N interacting particles distributed on a line of length L with Hamiltonian

$$\mathcal{H}(x) := \sum_{i=1}^{N} \frac{p_i^2}{2m} + \frac{g^2}{2} \sum_{\substack{i,j=1 \\ i \neq j}}^{N} \frac{1}{(q_i - q_j)^2}, \qquad g \in \mathbb{R}.$$

Here the phase space of the system is $\Omega := ([0, L] \times \mathbb{R})^N$.

3. Consider a system of N identical particles with mass $m > 0$ distributed on the surface of a sphere of radius 1 and subject to the gravitational potential energy. In spherical coordinates we can express the Hamiltonian as

$$\mathcal{H}(x) := \sum_{i=1}^{N} \left(\frac{1}{2m} \left(p_{\theta_i}^2 + \frac{p_{\phi_i}^2}{\sin^2 \theta_i} \right) + m g \cos \theta_i \right), \qquad g > 0.$$

Here the phase space of the system is $\Omega := ([0, \pi) \times [0, 2\pi) \times \mathbb{R}^2)^N$.

▶ Remarks:

- Note that the 2-body assumption on the interaction potential energy in (3.1) is quite restrictive and not really necessary. In principle, one can assume many-body interactions of the following type:

$$\mathcal{U}_{\text{tot}}(q_1, \ldots, q_N) := \sum_{k \geqslant 2} \sum_{1 \leqslant i_1 < \cdots < i_k \leqslant N} \mathcal{U}_k \left(q_{i_1}, \ldots, q_{i_k} \right), \qquad (3.3)$$

  where the functions $\mathcal{U}_k$ are taken to be invariant under permutation of their k arguments and invariant under translations. Then the interaction potential (3.3) coincides with the 2-body interaction in (3.1) if $k = 2$.

- Common assumptions on the 2-body interaction potential energy appearing in (3.1) are:

  1. $\mathcal{U}$ has a central symmetry, i.e., it depends only on the norms $\|q_i - q_j\|$ (*central potential*).

  2. $\mathcal{U}$ rapidly decreases when each distance $\|q_i - q_j\|$ increases (*short range potential*). In particular, if there exists $R > 0$ such that $\mathcal{U}(q_i - q_j) = 0$ if $\|q_i - q_j\| \geqslant R$, for all $i, j = 1, \ldots, N$, then $\mathcal{U}$ is a *finite range potential*.

  For the moment we do not make any assumption on the potential energies.

- It is natural to expect that if all potential energies in (3.1) vanish, or, more generally, are negligible, we should recover the free ideal gas described by the Maxwell distribution (2.23), leading to the *free ideal gas law*

$$P V = N \kappa T. \qquad (3.4)$$

  If the interaction potential energies in (3.1) are not negligible then we have a *real gas*, whose equation of state is not (3.4).

▶ The task of statistical mechanics is evidently not to follow the trajectories in the phase space Ω, which is impossible for many reasons (knowledge of x_0, complexity of Φ_t, etc.), but rather to derive the macroscopic properties from the laws governing the behavior of individual particles. The macroscopic properties are expressed in terms of thermodynamic variables. To summarize, the task is to construct *macroscopic states* from *microscopic states* in such a way that macroscopic states obey to those thermodynamic laws which are physically observed.

▶ The construction of thermodynamic quantities is done in terms of averaging operations. At this stage we have two fundamental problems:

1. The justification for the interpretation of averages as physical macroscopic quantities.

2. The development of methods to compute such averages, typically via asymptotic expansions reproducing physical thermodynamic quantities. This is done by considering limiting procedures which involve the number of degrees of freedom tending to infinity.

3.2 Definition of Gibbs ensemble

▶ The Gibbs formalism of statistical mechanics is based on the following construction, which was proposed for the first time by Boltzmann.

- *Microstates and macrostates.* A *microstate* of the system is defined by a point in the phase space Ω. A *macrostate* of the system corresponds to a set of microstates, denoted by $\mathcal{E} \subset \Omega$, which have the property to generate some prescribed thermodynamic properties. For example, if we interchange two particles we obtain a new point in Ω, but this does not change the macrostate. Therefore it is possible to consider $\mathcal{E}$ as being produced by an extremely numerous collection of kinematic states of the system in the same situation of thermodynamic equilibrium.

- *Distribution function on $\mathcal{E}$.* In general, points of $\mathcal{E}$ are not uniformly distributed. After a limiting procedure, we can regard $\mathcal{E}$ as a continuous set equipped with a *distribution function* $\rho : \Omega \to \mathbb{R}_+$, integrable over $\mathcal{E}$ (w.r.t. the Lebesgue measure). The pair $(\mathcal{E}, \rho)$ is called *Gibbs ensemble* and the space $\mathcal{E}$ is the *support of the distribution*.

  (a) The number of states $\nu(\mathcal{E}_0)$ contained in a measurable region $\mathcal{E}_0 \subset \mathcal{E}$ is given by the integral

  $$\nu(\mathcal{E}_0) := \int_{\mathcal{E}_0} \rho(x)\,dx. \tag{3.5}$$

  (b) Since ρ is not necessarily normalized to one, one can introduce a *probability distribution function* by defining

  $$\widetilde{\rho}(x) := \frac{\rho(x)}{\mathcal{Z}}, \qquad \mathcal{Z} := \int_{\mathcal{E}} \rho(x)\,dx.$$

Here $\mathcal{Z}$ is a function of the parameters of the system and is called *partition function* of $(\mathcal{E}, \rho)$. We will see that $\mathcal{Z}$ encodes almost everything about the thermodynamics of the system. The partition function describes how the probabilities are partitioned among different microstates. The letter $\mathcal{Z}$ stands for the German word *Zustandssumme*, "sum over states".

(c) If $f : \Omega \to \mathbb{R}$ is a function representing a measurable physical quantity at thermodynamic equilibrium, then its *ensemble average* w.r.t. ρ is given by

$$\langle f(x) \rangle_\rho := \frac{1}{\mathcal{Z}} \int_{\mathcal{E}} f(x)\,\rho(x)\,\mathrm{d}x, \tag{3.6}$$

where it is assumed that the integral converges.

▶ It is natural to identify the formalism of Gibbs ensembles with the formalism of measure spaces defined in Chapter 1. Indeed, we can write

$$(X, \mathscr{A}, \mu) \equiv (\mathcal{E}, \mathscr{B}(\mathcal{E}), \nu),$$

where ν is defined by (3.5) for any $\mathcal{E}_0 \in \mathscr{B}(\mathcal{E})$. In this language any function for which we compute (3.6) is a random variable (if the distribution is normalized to one).

▶ The above definition of Gibbs ensemble raises some difficult and intimately related problems:

1. *Existence and uniqueness of ensembles.* Given a classical system with many degrees of freedom, both existence and uniqueness of a Gibbs ensemble $(\mathcal{E}, \rho)$ which allows us to describe its thermodynamic behavior are not obvious.

2. *Ergodic problem.* Even less obvious is the interpretation of $\langle f(x) \rangle_\rho$ as the value attained by f at thermodynamic equilibrium. Indeed, it would be more natural to expect that the (uncomputable!) *time average* of f,

$$\langle f(x) \rangle_\infty := \lim_{T \to +\infty} \frac{1}{T} \int_0^T f(\Phi_t(x))\,\mathrm{d}t,$$

be the correct value attained at thermodynamic equilibrium. This dilemma is "solved" by the *ergodic hypothesis*, which says that $\langle f(x) \rangle_\rho = \langle f(x) \rangle_\infty$ (almost) everywhere.

3. *Orthodicity of ensembles.* Even if $(\mathcal{E}, \rho)$ exists and is a well-defined mathematical object, then it is not guaranteed that it allows us to derive a physically correct thermodynamic behavior. If this happens, then $(\mathcal{E}, \rho)$ is called *orthodic*.

4. *Equivalence of ensembles.* If two distinct orthodic ensembles do exist, then they must be equivalent in some sense. This difficult problem invokes the notion of the *thermodynamic limit* (TL), whose existence, on the other hand, guarantees that thermodynamic potentials are extensive. The formal definition of the TL depends on the boundary conditions. Loosely speaking, the TL of a system is the limit for a large number N of particles where the volume is taken to grow in proportion with the number of particles. In other words, the TL can be interpreted as the limit of a system with a large volume, with the particle density held constant. In this limit, the macroscopic thermodynamics is recovered, i.e., the Gibbs ensemble used to describe the system is orthodic.

▶ We will see that in the classical development of statistical mechanics for continuous systems there are at least three distinct classical ensembles. On the one hand, it is quite easy to understand to which physical situations they correspond. On the other hand, to prove in a rigorous way under which conditions they are orthodic and equivalent is a much more complicated task. Here are the three physical situations (more precisely three different *boundary conditions*) we are going to consider. The basic model is given by a (theoretical) gas of particles governed by a Hamiltonian of type (3.1). Then there are the following cases.

(ME) The system is *closed*, i.e., N is constant, and *isolated*, i.e., the value E attained by the Hamiltonian is constant.

- The *microcanonical ensemble* (ME) is the ensemble that is used to represent the possible states of such system.
- The system cannot exchange energy or particles with its environment, so that (by conservation of energy) the energy of the system remains exactly known as time goes on.
- The thermodynamic variables are the total number of particles N, the volume V and the total energy E. The thermodynamic potential of the ME is the *entropy*, which is a macroscopic extensive function defined by

$$S(N, V, E) := \kappa \log \mathcal{Z}_{\mathrm{M}}(N, V, E),$$

where $\mathcal{Z}_{\mathrm{M}}(N, V, E)$ is called *microcanonical partition function*. It encodes how the probabilities are partitioned among different microstates and it allows to derive in a systematic way the thermodynamics of the system.

(CE) The system is closed and maintained in thermal contact with a thermostat (also called heat bath) at fixed temperature T. Thermal contact means that the system can exchange energy through an interaction which must be weak as to not significantly perturb the microstates of the system.

- The *canonical ensemble* (CE) is the ensemble that is used to represent the possible states of such system.

- The system can exchange energy with the thermostat, so that various possible states of the system can differ in total energy.
- The thermodynamic variables are the total number of particles N, the volume V and the temperature T. The thermodynamic potential of the CE is the *free energy* (or *Helmholtz energy*), which is a macroscopic extensive function defined by

$$F(N, V, T) := -\kappa\, T \log \mathcal{Z}_{\mathrm{C}}(N, V, T),$$

where $\mathcal{Z}_{\mathrm{C}}(N, V, T)$ is called *canonical partition function*.

(GE) The system is neither closed nor isolated and is maintained in thermodynamic equilibrium with a reservoir.

- The *grand canonical ensemble* (GE) is the ensemble that is used to represent the possible states of such system.
- The system can exchange energy and particles with the reservoir, so that various possible states of the system can differ in both their total energy and total number of particles.
- The thermodynamic variables are the chemical potential μ, the volume V and the temperature T. The thermodynamic potential of the GE is the *grand potential* (or *Landau potential*), which is a macroscopic extensive function defined by

$$O(\mu, V, T) := -\kappa\, T \log \mathcal{Z}_{\mathrm{G}}(\mu, V, T),$$

where $\mathcal{Z}_{\mathrm{G}}(N, V, T)$ is called *grand canonical partition function*.

▶ Remarks:

- The ME does not correspond to any experimentally realistic situation. With a real physical system there is at least some uncertainty in energy, due to uncontrolled factors in the preparation of the system. Besides the difficulty of finding an experimental analogue, it is difficult to carry out calculations that satisfy exactly the requirement of fixed energy since it prevents logically independent parts of the system from being analyzed separately.

- The typical realization of the CE is the one where the system with Hamiltonian (3.1) is in contact with a second much larger system, called thermostat, described by a Hamiltonian $\mathscr{H}_{\mathrm{ther}}$. The thermostat is needed to fix the temperature T and the energy of the first system fluctuates near a prescribed average. The total system has Hamiltonian

$$\mathscr{H}_{\mathrm{tot}} = \mathscr{H} + \mathscr{H}_{\mathrm{ther}} + \mathscr{H}_{\mathrm{c}},$$

where $\mathscr{H}_{\mathrm{c}}$ is a coupling term which generates weak random perturbations. Then the total system is closed and isolated.

▶ In the three above listed cases the problem will be to give a mathematical formulation of the pair $(\mathcal{E}, \rho)$ and then to derive the thermodynamics.

3.2.1 The ergodic hypothesis

▶ Before going into the details of the three ensembles we discuss the justification for Gibbs' interpretation of observable quantities as ensemble averages (instead of time averages). This will raise a crucial problem, called *ergodic problem*, and will help us in understanding the mathematical structure that the space $\mathcal{E}$ must have.

▶ To fix ideas we will always assume that our system is closed and isolated. Furthermore it is always assumed that the system under consideration has a time evolution described by the canonical Hamiltonian flow $\Phi_t : \mathbb{R} \times \Omega \to \Omega$ generated by a Hamiltonian of type (3.1).

- Define the $(2\ell N - 1)$-dimensional invariant manifold given by the level set of the total energy:
$$\Sigma_E := \{x \in \Omega \, : \, \mathcal{H}(x) = E\},$$
  where $E \geqslant 0$ is fixed. We assume that Σ_E encloses a compact region of Ω.

  (a) Invariance of Σ_E means that $\Phi_t(\Sigma_E) = \Sigma_E$ for all $t \in \mathbb{R}$.
  (b) The Hamiltonian vector field is tangent to Σ_E at every point, while the vector field $\mathrm{grad}_x \mathcal{H}(x)$ is orthogonal to Σ_E at every point.
  (c) The invariant infinitesimal $\mathrm{d}x$ can be written as $\mathrm{d}x = \mathrm{d}\sigma \, \mathrm{d}n$, where $\mathrm{d}\sigma$ is the infinitesimal surface element of Σ_E and $\mathrm{d}n = \mathrm{d}\mathcal{H} / \| \mathrm{grad}_x \mathcal{H}(x)\|$ is the infinitesimal normal vector to Σ_E. Then we write

$$\mathrm{d}x = \mathrm{d}\Sigma_E \, \mathrm{d}\mathcal{H}, \qquad \mathrm{d}\Sigma_E := \frac{\mathrm{d}\sigma}{\| \mathrm{grad}_x \mathcal{H}(x)\|}. \tag{3.7}$$

  Note that $\mathrm{d}\Sigma_E$ is invariant under Φ_t since $\mathrm{d}\mathcal{H}$ is. In particular, the area of Σ_E can be expressed as

$$\mathrm{Area}(\Sigma_E) = \int_{\Sigma_E} \mathrm{d}\Sigma_E = \int_{\Sigma_E} \frac{\mathrm{d}\sigma}{\| \mathrm{grad}_x \mathcal{H}(x)\|}. \tag{3.8}$$

- We claim that $\mathcal{E} = \Sigma_E$, since Σ_E is exactly the manifold defined by all microstates corresponding to the fixed value E, corresponding to a well-defined macrostate.

- If x and y belong to the same trajectory in Ω, i.e., $x = \Phi_t(y)$ for some t, then there exists between them a deterministic correspondence, and therefore we must attribute to the two points the same probability density. This proves that the density ρ is an integral of motion like $\mathcal{H}$, i.e.,

$$\rho\left(\Phi_t(x)\right) = \rho(x) \qquad \forall x \in \Omega, t \in \mathbb{R}. \tag{3.9}$$

Example 3.2 (*Integrals over* Ω)

Consider integrals of the form

$$I(E) := \int_{\Omega_E} f(x)\, dx,$$

where $f : \Omega \to \mathbb{R}$ is a measurable function and

$$\Omega_E := \{x \in \Omega : 0 \leqslant \mathscr{H}(x) \leqslant E\},$$

where $E \geqslant 0$ is fixed. Note that Ω_E is the (compact) region of Ω which is enclosed by Σ_E. We can think of Ω_E as a foliated manifold whose leaves are submanifolds $\Sigma_{\mathscr{H}}$ with $0 \leqslant \mathscr{H} \leqslant E$.

- In view of (3.7) we have

$$I(E) = \int_0^E d\mathscr{H} \int_{\Sigma_{\mathscr{H}}} f(x)\, d\Sigma_{\mathscr{H}}.$$

- We also have

$$\int_{\Sigma_E} f(x)\, d\Sigma_E = \frac{\partial I}{\partial E},$$

which implies ($f \equiv 1$)

$$\mathrm{Area}(\Sigma_E) := \int_{\Sigma_E} d\Sigma_E = \frac{\partial}{\partial E} \int_{\Omega_E} dx =: \frac{\partial}{\partial E} \mathrm{Vol}(\Omega_E).$$

▶ To understand the structure that $(\mathcal{E}, \rho)$ must have and to justify Gibbs's assumption on the averages we need some preliminary results and notions. Let $\Omega_0 \subset \Omega$ be an arbitrary Lebesgue measurable subset and $\Omega_0(t) := \Phi_t(\Omega_0)$ be its image under the Hamiltonian flow at time t.

- Let $f : \Omega_0 \to \mathbb{R}$ be an integrable function. Then we have

$$\int_{\Omega_0} f(x)\, dx = \int_{\Omega_0(t)} f\left(\Phi_{-t}(y)\right) dy, \qquad y := \Phi_t(x). \tag{3.10}$$

- Ω_0 is *invariant* under Φ_t if $\Omega_0(t) = \Omega_0$ for all $t \in \mathbb{R}$. Equivalently, Ω_0 is invariant if

$$\int_{\Omega_0} f\left(\Phi_t(x)\right) dx = \mathrm{const.}$$

for every integrable function $f : \Omega_0 \to \mathbb{R}$.

- Let $\rho : \Omega \to \mathbb{R}_+$ be a distribution function on Ω. Then we define the *ρ-measure* of Ω_0 by

$$|\Omega_0|_\rho := \int_{\Omega_0} \rho(x)\, dx, \tag{3.11}$$

where the integral is assumed to be convergent.

(a) Any property that is satisfied everywhere except in a set of ρ-measure zero is said to hold *ρ-almost everywhere*.

(b) A function $f : \Omega_0 \to \mathbb{R}$ is *ρ-integrable* if and only if

$$\int_{\Omega_0} |f(x)| \rho(x)\, dx < +\infty.$$

(c) Ω_0 is *metrically indecomposable* w.r.t. $|\cdot|_\rho$ if it cannot be decomposed into the union of two disjoint ρ-measurable subsets.

▶ We have the following statement.

Theorem 3.1

The *ρ-measure is invariant under Φ_t, i.e.,*

$$|\Omega_0(t)|_\rho = |\Omega_0|_\rho \qquad \forall\, t \in \mathbb{R}$$

for any measurable subset $\Omega_0 \subset \Omega$.

Proof. From formulas (3.9), (3.10) and (3.11) we have:

$$
\begin{aligned}
|\Omega_0|_\rho &:= \int_{\Omega_0} \rho(x)\, dx = \int_{\Omega_0(t)} \rho\left(\Phi_{-t}(y)\right) dy \\
&= \int_{\Omega_0(t)} \rho(y)\, dy =: |\Omega_0(t)|_\rho,
\end{aligned}
$$

where $y := \Phi_t(x)$. ■

▶ Remarks:

- Consider the map $\Phi := \Phi_t$, with t fixed, and denote by $\mathscr{B}(\mathcal{E})$ the σ-algebra of Borel sets on $\mathcal{E}$. The system $(\mathcal{E}, \mathscr{B}(\mathcal{E}), \rho, \Phi)$ is a measurable dynamical system.

- The measure $|\Omega_0|_\rho$ is proportional (equal if ρ is a probability density) to the probability that the system is in a microscopic state described by a point in the space Ω belonging to Ω_0. In particular,

$$|\mathcal{E}|_\rho = \int_{\mathcal{E}} \rho(x)\, dx = \mathcal{Z}.$$

▶ We now consider two problems:

1. *Existence of time average of a measurable function f.* The experimental observation of a macroscopic quantity, represented by a function f, is not done by selecting a precise microscopic state, i.e., a point in Ω, but rather it refers to an arc of the trajectory of a point in the space Ω. It seems close to the reality of a measurement process to consider the time average of f on arcs of the trajectory of the system.

2. *Identification of time average with ensemble average.* Even if the time average does exist, its actual computation is only a hypothetical operation, as neither it is possible to determine a Hamiltonian flow of such complexity nor to know its initial conditions.

▶ The first problem is solved by next claim.

Theorem 3.2 (*Birkhoff*)

Let $\Omega_0 \subset \Omega$ be a subset with finite Lebesgue measure invariant w.r.t. a Hamiltonian flow Φ_t. Let $f : \Omega_0 \to \mathbb{R}$ be an integrable function. Then the following claims are true.

1. *The limit*

$$\langle f(x) \rangle_\infty := \lim_{T \to \pm\infty} \frac{1}{T} \int_0^T f\left(\Phi_t(x)\right) dt \tag{3.12}$$

exists for almost every $x \in \Omega_0$ (w.r.t. the Lebesgue measure).

2. *There holds*

$$\langle f\left(\Phi_t(x)\right) \rangle_\infty = \langle f(x) \rangle_\infty$$

for almost every $x \in \Omega_0$.

No Proof.

▶ Remarks:

- The limit (3.12) defines the *time average* of f along the flow Φ_t on an infinite time interval. In principle, for finite and different time intervals, it can take very different values. Theorem 3.2 guarantees the existence, for almost every trajectory, of the time average, and it establishes that averages over sufficiently long intervals are approximately equal (as they must all tend to the value $\langle f(x) \rangle_\infty$).

- We are neglecting a critical discussion on the identification of the result of a measurement with the time average. In principle, one should consider the following problem: how much time must pass (hence how large must T be in (3.12)) for the difference between the average of a quantity f on the interval $[0, T]$ and the time average (3.12) to be less than a prescribed tolerance? This problem is known as the *problem of relaxation times* at the equilibrium value for an observable quantity. It is a problem of central importance in classical statistical mechanics, and it is still the object of intensive research.

▶ The second problem is solved *a priori* by the celebrated *ergodic hypothesis*.

Ergodic hypothesis (Boltzmann)

> If Φ_t visits every subset of $\mathcal{E}$ with positive measure, then the time average of a function $f : \mathcal{E} \to \mathbb{R}$ can be identified with its ensemble average, i.e.,
>
> $$\langle f(x) \rangle_\infty = \langle f(x) \rangle_\rho , \qquad (3.13)$$
>
> for ρ-almost every $x \in \mathcal{E}$.

▶ Remarks:

- Later we will try to clarify under which conditions the identification (3.13) can be done.

- To understand what is the degree of confidence we may attach to $\langle f(x) \rangle_\rho$ as the equilibrium value of an observable we can compute the *mean quadratic fluctuation* of f:

$$\mathrm{mqf}(f(x)) := \frac{\langle f^2(x) \rangle_\rho - \langle f(x) \rangle_\rho^2}{\langle f(x) \rangle_\rho^2}.$$

  Typically we will consider extensive quantities f for which the ensemble averages $\langle f^2(x) \rangle_\rho$ and $\langle f(x) \rangle_\rho^2$ are both of order N^2. Hence what is required for $\langle f(x) \rangle_\rho$ to be a significant value is that $\mathrm{mqf}(f(x)) \ll 1$ for $N \gg 1$.

▶ In the classical literature, the fact that the ergodic hypothesis has been formulated by Boltzmann assuming the existence of a trajectory passing through all points in the phase space accessible to the system (hence corresponding to a fixed value of the energy) is often discussed. Clearly this condition would be sufficient to ensure that temporal averages and ensemble averages are interchangeable, but at the same time its impossibility is evident. Indeed, the phase trajectory of a Hamiltonian flow has measure zero since it is the image of a regular curve, and hence it can be dense at most on the constant energy surface.

▶ The reasoning of Boltzmann is, however, much richer and more complex (and maybe this is the reason why it was not appreciated by his contemporaries) and it deserves a brief discussion which we take from " Statistical Mechanics: a Short Treatise" by G. Gallavotti.

- Consider a theoretical gas of particles, closed and isolated, described by the Hamiltonian (3.1).

- Instead of assuming that the system can take a continuum of states in the space Ω, we decompose the latter into n small identical cells Ω_i, each determining the position and momentum of each particle with the uncertainty unavoidable in every measurement process.

- If $h > 0$ denotes the uncertainty in the measurements of position and momentum, and hence if $\delta p \, \delta q \approx h$ (here δ denotes a small variation), then h^{3N} is the volume of a cell. The microscopic state space is then the set of the cells Ω_i partitioning Ω.

- The Hamiltonian flow induces a transformation $\Phi := \Phi_\tau$ which transforms each cell Ω_i into another cell: $\Phi : \Omega_i \mapsto \Omega_j$, $i \neq j$. Here τ is a "microscopic time", very short w.r.t. the duration of any macroscopic measurement of the system and on a scale in which the motion of the particles can be measured.

- The map Φ is an injective surjective function. Indeed Φ is the canonical linear map obtained by solving over a time interval τ the canonical Hamiltonian equations linearized at the centre of the cell Ω_i considered. The effect of Φ is therefore to permute the cells Ω_i among them.

- Since the system is closed and isolated, its energy E is macroscopically fixed (and lies between E and $E + \delta E$, with δE "macroscopically small"). Since the volume accessible to the particles is finite, the number n of cells representing the energetically possible states is very large, but finite.

- Given a measurable function defined on the phase space, the ergodic hypothesis now states that its time average and its ensemble average coincide. This is equivalent to assume that Φ acts as a *one-cycle permutation*: *a given cell Ω_i evolves successively into different cells until it returns to the initial state in a number of steps equal to the number n of cells. It follows that, by numbering the cells appropriately, we have*

$$\Phi(\Omega_i) = \Omega_{i+1}, \qquad i = 1, \ldots, n,$$

  with the periodic condition $\Omega_{n+1} = \Omega_1$. In other words as time evolves every cell evolves, visiting successively all other cells with equal energy.

- Even if not strictly true the above claim should hold at least for the purpose of computing the time averages of the observables relevant for the macroscopic properties of the system. The basis for such a celebrated (and much criticized) hypothesis rests on its conceptual simplicity: it says that *all cells with the same energy are equivalent.*

▶ There are cases (already well known to Boltzmann) in which the hypothesis is manifestly false.

- For instance, if the system is enclosed in a perfectly spherical container then the evolution keeps the angular momentum constant. Hence cells with a different total angular momentum cannot evolve into each other.

- This extends to the cases in which the system admits other integrals of motion, besides the energy, because the evolving cells must keep the integrals of motion equal to their initial values. This means that the existence of other integrals of motion besides the energy is, essentially, the most general case in which the ergodic hypothesis fails: in fact when the evolution is not a one-cycle permutation of the phase space cells with given energy, then one can decompose it into cycles. One can correspondingly define a function f by associating with each cell of the same cycle the very same (arbitrarily chosen) value of f, different from that of cells of any other cycle. Obviously the function f so defined is an integral of motion that can play the same role as the angular momentum in the previous example.

- Thus, if the ergodic hypothesis failed to be verified, then the system would be subject to other conservation laws, besides that of the energy. In such cases it would be natural to imagine that *all* the integrals of motion were fixed and to ask oneself which are the qualitative properties of the motions with energy E, when all the other integrals of motion are also fixed. Clearly in this situation the motion will be *by construction* a simple cyclic permutation of all the cells compatible with the prefixed energy and other constants of motion values.

- Hence it would be more convenient to define formally the notion of *ergodic probability distribution* on Ω saying that *a set of phase space cells is ergodic if Φ maps it into itself and if Φ acting on the set of cells is a one-cycle permutation of them.*

- Therefore, in some sense, the ergodic hypothesis would not be restrictive and it would simply become the statement that one studies the motion after having *a priori* fixed all the values of the integrals of motion.

3.2.2 The problem of existence of integrals of motion

▶ From the previous discussion of the ergodic hypothesis we can argue that such assumption must admit some deep Hamiltonian formulation which highlights the role of integrals of motion of the system. Indeed, an interesting branch of Hamiltonian mechanics is devoted to this problem. We just want to understand the main results of this theory, without going into the details of Hamiltonian perturbation theory.

▶ We continue to assume that our system is closed and isolated and governed by the Hamiltonian (3.1). The following claim holds true.

Theorem 3.3

Let $(\mathcal{E}, \rho)$ be a Gibbs ensemble.

1. $(\mathcal{E}, \rho)$ is metrically indecomposable w.r.t. $|\cdot|_\rho$ if and only if the time average of any ρ-integrable function $f : \mathcal{E} \to \mathbb{R}$ is constant for ρ-almost every $x \in \mathcal{E}$.

2. If $(\mathcal{E}, \rho)$ is metrically indecomposable w.r.t. $|\cdot|_\rho$ and $f : \mathcal{E} \to \mathbb{R}$ is a ρ-integrable function, then

$$\langle f(x) \rangle_\infty = \langle f(x) \rangle_\rho \tag{3.14}$$

for ρ-almost every $x \in \mathcal{E}$.

Proof. We prove only the second claim. We proceed by steps.

- Let $\alpha \in \mathbb{R}$ be the value of the time average of f, i.e., $\langle f(x) \rangle_\infty = \alpha$. Fix $T > 0$ and consider the identity

$$
\begin{aligned}
\alpha \;=\; & \frac{1}{|\mathcal{E}|_\rho} \int_\mathcal{E} \left(\alpha - \frac{1}{T} \int_0^T f\left(\Phi_t(x)\right) dt \right) \rho(x)\, dx \\
& + \frac{1}{|\mathcal{E}|_\rho} \int_\mathcal{E} \left(\frac{1}{T} \int_0^T f\left(\Phi_t(x)\right) dt \right) \rho(x)\, dx.
\end{aligned}
$$

- Both $\mathcal{E}$ and ρ are invariant under the action of Φ_t. Hence we have

$$
\begin{aligned}
\frac{1}{|\mathcal{E}|_\rho} \int_\mathcal{E} \left(\frac{1}{T} \int_0^T f\left(\Phi_t(x)\right) dt \right) \rho(x)\, dx \;=\; & \frac{1}{T|\mathcal{E}|_\rho} \int_0^T dt \int_\mathcal{E} f(x) \rho(x)\, dx \\
=\; & \langle f(x) \rangle_\rho.
\end{aligned}
$$

- Therefore,

$$\alpha - \langle f(x) \rangle_\rho = \frac{1}{|\mathcal{E}|_\rho} \int_\mathcal{E} (\alpha - f_T(x)) \rho(x)\, dx,$$

where

$$f_T(x) := \frac{1}{T} \int_0^T f\left(\Phi_t(x)\right) dt.$$

- For any arbitrary $\varepsilon > 0$ define the sets

$$\mathcal{E}_T^{(1)} := \{ x \in \mathcal{E} : |\alpha - f_T(x)| < \varepsilon \}, \qquad \mathcal{E}_T^{(2)} := \mathcal{E} \setminus \mathcal{E}_T^{(1)}.$$

- Then we have:

$$\left| \int_\mathcal{E} (\alpha - f_T(x)) \rho(x)\, dx \right| \leqslant \varepsilon \left| \mathcal{E}_T^{(1)} \right|_\rho + |\alpha| \left| \mathcal{E}_T^{(2)} \right|_\rho + \int_{\mathcal{E}_T^{(2)}} |f_T(x)| \rho(x)\, dx.$$

- Since $f_T(x) \to \alpha$ as $T \to +\infty$ for ρ-almost every $x \in \mathcal{E}$, then $|\mathcal{E}_T^{(2)}|_\rho \to 0$ as $T \to +\infty$. Hence, for T sufficiently large and $\varepsilon' > 0$ we have $|\mathcal{E}_T^{(2)}|_\rho < \varepsilon'$ and $|\Phi_t(\mathcal{E}_T^{(2)})|_\rho < \varepsilon'$ for all $t \in [0, T]$ thanks to Theorem 3.1. Thanks to the absolute continuity of the integral we can choose ε' sufficiently small (namely T sufficiently large) in such a way that

$$\int_{\mathcal{E}_T^{(2)}} |f_T(x)| \, \rho(x) \, dx \leqslant \frac{1}{T} \int_0^T dt \int_{\Phi_t\left(\mathcal{E}_T^{(2)}\right)} |f(x)| \, \rho(x) \, dx < \varepsilon.$$

- Therefore we have

$$\left| \int_{\mathcal{E}} (\alpha - f_T(x)) \, \rho(x) \, dx \right| \leqslant \varepsilon \left| \mathcal{E}_T^{(1)} \right|_\rho + |\alpha| \varepsilon + \varepsilon.$$

Since ε is arbitrary we get $\langle f(x) \rangle_\rho = \alpha = \langle f(x) \rangle_\infty$ as desired.

The second claim is proved. ∎

▶ We now give a more formal definition of the ergodic hypothesis. It is the basis of the Gibbsian formalism of continuous statistical systems, as it allows one to interpret the averages of observable thermodynamic quantities as their equilibrium values.

Definition 3.1

1. $(\mathcal{E}, \rho)$ is **ergodic** if and only if condition (3.14) is satisfied for any ρ-integrable function $f : \mathcal{E} \to \mathbb{R}$.

2. If a Hamiltonian system admits an ergodic ensemble $(\mathcal{E}, \rho)$, then we say that it satisfies the **ergodic hypothesis**.

▶ The next claim gives an alternative formulation of ergodicity of a Gibbs ensemble in terms of existence of integrals of motion of the system.

Theorem 3.4

$(\mathcal{E}, \rho)$ is ergodic if and only if $\mathscr{H}$ and ρ are the only integrals of motion.

Proof. We prove the claim only in one direction. Assume that our system admits a third measurable integral of motion f functionally independent on ρ. Then we would have a family of invariant level sets

$$\Sigma_c := \{x \in \Omega : f(x) = c\},$$

with $c \in \mathbb{R}$. Any subset

$$\Sigma_{c_1, c_2} := \bigcup_{c_1 \leqslant c \leqslant c_2} \Sigma_c$$

is also invariant and of positive Lebesgue measure for a proper choice of $c_1, c_2 \in \mathbb{R}$. Then, since $\mathcal{E} = \Sigma_E$ corresponds to an isolated system, the intersection $\mathcal{E} \cap \Sigma_{c_1, c_2}$ is another invariant set. For a proper choice of $c_1, c_2 \in \mathbb{R}$, since f and ρ are functionally independent integrals of motion we have

$$|\mathcal{E}|_\rho > |\mathcal{E} \cap \Sigma_{c_1,c_2}|_\rho > 0,$$

which implies that $\mathcal{E}$ is metrically decomposable w.r.t. $|\cdot|_\rho$. Thus it is not ergodic. ∎

▶ From Theorem 3.4 there follows immediately the next claim.

Corollary 3.1

> If $(\mathcal{E}, \rho)$ is ergodic then ρ is an integral of motion functionally dependent on $\mathcal{H}$, i.e., $\rho = \rho(\mathcal{H})$.

▶ From Theorem 3.4 and its Corollary 3.1 we immediately deduce that a completely integrable Hamiltonian system is not ergodic. For systems which are typically studied within the framework of statistical mechanics, it is possible, in general, to recognize in the Hamiltonian (3.1) a part corresponding to a completely integrable system.

- Tipically, the difference between the Hamiltonian (3.1) and an integrable Hamiltonian is small, and the system is therefore in the form of a *quasi-integrable system*. In other words the Hamiltonian (3.1) can be written in the form

$$\mathcal{H}(x, \varepsilon) = \mathcal{H}_0(x) + \varepsilon \mathcal{H}_{\text{per}}(x), \qquad |\varepsilon| \ll 1. \tag{3.15}$$

  Here $\mathcal{H}_0$ is the Hamiltonian of a completely canonically integrable system and $\mathcal{H}_{\text{per}}$ is a regular perturbation (typically the potential energy).

- Under some quite relaxed conditions, Hamiltonian perturbation theory guarantees the existence of a symplectic transformation from x to *action-angle variables* $(J, \chi) \in \mathbb{R}^L \times \mathbb{T}^L$ (L is the total number of degrees of freedom) such that the Hamiltonian (3.15) can be written as

$$\mathcal{H}(J, \chi, \varepsilon) = \mathcal{H}_0(J) + \varepsilon \mathcal{H}_{\text{per}}(J, \chi). \tag{3.16}$$

  Note that the integrable Hamiltonian $\mathcal{H}_0$ is expressed in terms of the action variables J, namely in terms of integrals of motion.

- Note that the possibility for (3.16) to be ergodic lies in the perturbation. Indeed, if $\varepsilon = 0$ the Hamiltonian (3.16) reduces to the integrable Hamiltonian $\mathcal{H}_0$, which depends only on the action variables J. The L action variables are nothing but L functionally independent and Poisson involutive integrals of motion. From the topological point of view they induce a *foliation* of the phase space in invariant tori, namely metric decomposability.

- It was proved by Poincaré that *under suitable regularity, genericity and non-degeneracy assumptions (actually satisfied by many systems of interest for statistical mechanics) there do not exist integrals of motion which are regular in ε, J, χ and functionally independent on the Hamiltonian (3.16)*. We claim a generalization of the last statement, due to E. Fermi: *under the same regularity, genericity and non-degeneracy assumptions of the theorem on non-existence of integrals of motion by Poincaré, a quasi-integrable Hamiltonian system (3.15) with $L > 2$ degrees of freedom does not admit invariant submanifolds with dimension $2L - 1$ and regular dependence on ε which are different from the energy level sets.*

- As a matter of fact, we do not know any physical system described by a Hamiltonian such as (3.1) (or (3.16)), where the potential energy is a regular function of its arguments (excluding therefore the possibility of situations such as that of a hard sphere gas with perfectly elastic collisions), for which the ergodic hypothesis has been proved! The problem of the ergodicity of Hamiltonian systems is therefore still fundamentally open, and is the object of intensive research.

Example 3.3 (*The harmonic oscillator*)

Consider the Hamiltonian of a one-dimensional harmonic oscillator:

$$\mathcal{H}(q, p) := \frac{1}{2} \left(\frac{p^2}{m} + m\,\omega^2\,q^2 \right).$$

Such system has one degree of freedom and it is trivially integrable.

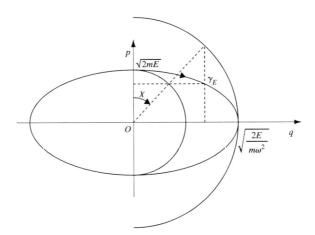

Fig. 3.1. Phase plane of the harmonic oscillator ([FaMa]).

- In the phase plane, the cycles γ_E of the energy level curve,

$$\frac{1}{2} \left(\frac{p^2}{m} + m\,\omega^2\,q^2 \right) = E, \qquad E > 0,$$

enclose the area $(2\pi/\omega)E$. By definition of action variable we get

$$J := \frac{1}{2\pi} \oint_{\gamma_E} p\, dq = \frac{1}{2\pi} \frac{2\pi}{\omega} E = \frac{E}{\omega}.$$

The meaning of the angle variable χ can be seen in the figure.

- It turns out the Hamiltonian written in terms of action-angle variables depends only on J and reads

$$\mathscr{H}(J) = \omega J.$$

- The symplectic transformation between (q, p) and $(J, \chi) \in \mathbb{R}_+ \times S^1$ is defined by

$$(q, p) = \left(\sqrt{\frac{2J}{m\omega}} \sin\chi, \sqrt{2m\omega J} \cos\chi \right).$$

▶ The Gibbs formalism of continuous systems is based on the ergodic hypothesis for the Hamiltonian (3.1). Here is a quite typical recipe used to derive the thermodynamic behavior for which (3.1) is responsible.

- The Hamiltonian (3.1) is written as

$$\mathscr{H}(x, \varepsilon) = \mathscr{H}_0(x) + \varepsilon \mathscr{H}_{\text{per}}(x), \qquad |\varepsilon| \ll 1,$$

where $\mathscr{H}_0$ is the Hamiltonian of a completely canonically integrable system and $\mathscr{H}_{\text{per}}$ is a perturbation (typically potential energies or a part of them).

- The ergodicity of the system is, by hypothesis, guaranteed by $\mathscr{H}_{\text{per}}$, but, in the derivation of thermodynamic quantities, the contribution of $\mathscr{H}_{\text{per}}$ is neglected.

Example 3.4 (*The free ideal gas*)

For a sufficiently diluted particle free gas with N ideal identical particles of mass m, the interaction potential $\mathscr{U}$ can be considered almost always as a small perturbation, because it can always be neglected except during collisions. Therefore the Hamiltonian (3.1) takes the form:

$$\mathscr{H}(x) := \mathscr{H}_0(p_1, \ldots, p_N) + \varepsilon \mathscr{H}_{\text{per}}(q_1, \ldots, q_N), \qquad |\varepsilon| \ll 1,$$

where the integrable part is

$$\mathscr{H}_0(p_1, \ldots, p_N) := \sum_{i=1}^{N} \frac{\|p_i\|^2}{2m},$$

and the perturbation is

$$\mathscr{H}_{\text{per}}(q_1, \ldots, q_N) := \sum_{1 \leqslant i < j \leqslant N} \mathscr{U}(q_i - q_j).$$

If $\varepsilon \mathscr{H}_{\text{per}}$ is negligible then it does not play any role in the derivation of the thermodynamics, which is completely determined by $\mathscr{H}_0$.

3.3   Microcanonical ensemble

▶ Let us recall the physical situation corresponding to the microcanonical ensemble (ME):

- The system, whose (ergodic) Hamiltonian is (3.1), is closed, i.e., N is constant, and isolated, i.e., the value E attained by the Hamiltonian is constant.

- The system cannot exchange energy or particles with its environment, so that the energy of the system remains exactly known as time goes on.

- The thermodynamic variables are the total number of particles N, the volume V and the total energy E.

As already mentioned, the definition of thermodynamic quantities by using the ME has only a theoretical interest, since the system itself is not accessible to any measurement.

▶ We already know that $\mathcal{E}_M = \Sigma_E$, where Σ_E is the $(2\ell N - 1)$-dimensional invariant manifold given by the level set of the total energy:

$$\Sigma_E := \{x \in \Omega : \mathcal{H}(x) = E\},$$

where $E \geqslant 0$ is fixed. It is always assumed that Σ_E encloses a compact region of Ω. Due to the fact that the system must be ergodic we are forced to claim that the equilibrium configuration must be described by a distribution function ρ_M that is constant on Σ_E.

▶ We give the following definition.

Definition 3.2

The **microcanonical ensemble** *is defined by*

$$(\mathcal{E}_M, \rho_M) := \left(\Sigma_E, \frac{\delta(\mathcal{H}(x) - E)}{N!} \right). \tag{3.17}$$

Here δ denotes the Dirac delta function.

▶ Remarks:

- The (one-dimensional) Dirac delta function is a generalized function, or *distribution*, on $\mathbb{R}$ that is zero everywhere except when its argument is zero, with an integral of one over $\mathbb{R}$. From a purely mathematical viewpoint, the Dirac delta function is not strictly a function, because any extended-real function that is equal to zero everywhere but a single point must have total integral zero. It only makes sense as a mathematical object when it appears inside an integral.

- As expected, ρ_M in (3.17) is defined in terms of the Hamiltonian. By using the Dirac delta function we get

$$\int_\Omega \delta(\mathcal{H}(x) - E) \, dx = \text{Area}(\Sigma_E),$$

which shows that the equilibrium distribution of the microscopic states is the uniform distribution on the energy level set Σ_E. In simple terms, the Dirac delta function selects out those configurations that have the same energy as the specified energy E.

- For physical reasons, the probability distribution in (3.17) should be corrected by an overall factor which adjusts the dimensions (dx is not dimensionless!). The origin of such correction, which is expressed in terms of the *Planck constant* $\hbar$, can be explained only in terms of quantum mechanics. We will neglect this correction setting $\hbar \equiv 1$. A drawback will be that arguments of logarithms, for instance defining the entropy, will carry physical dimensions, which is obviously incorrect.

- The factor $1/N!$ gives the correct counting of particles. Indeed particles are indistinguishable, so that any permutation of them gives rise to the same macroscopic state (*Boltzmann counting*). The indistinguishability of particles is a delicate concept and its understanding can be explained only in terms of quantum mechanics.

- The probability distribution function of the ME is

$$\tilde{\rho}_M(x) := \frac{1}{\mathcal{Z}_M(N,V,E)} \frac{\delta(\mathcal{H}(x) - E)}{N!},$$

where

$$\mathcal{Z}_M(N,V,E) := \frac{\mathrm{Area}(\Sigma_E)}{N!}$$

is the *microcanonical partition function*.

- The ME-average of a measurable function $f : \Omega \to \mathbb{R}$ is by definition (see (3.8))

$$
\begin{aligned}
\langle f(x) \rangle_M &:= \frac{1}{\mathcal{Z}_M(N,V,E)} \frac{1}{N!} \int_\Omega f(x)\, \delta(\mathcal{H}(x) - E)\, dx \\
&= \frac{1}{\mathcal{Z}_M(N,V,E)} \frac{1}{N!} \int_{\Sigma_E} \frac{f(x)}{\| \mathrm{grad}_x \mathcal{H}(x) \|}\, d\sigma,
\end{aligned}
$$

where $d\sigma$ is an infinitesimal element of the energy surface Σ_E.

- The definition of ρ_M (and of $\mathcal{Z}_M$) can be justified by a limiting procedure.

  (a) Introduce an approximation of the ensemble. Define $\mathcal{E}_{\delta E}$ as the the accessible part of Ω lying between the two shells Σ_E and $\Sigma_{E+\delta E}$, where $\delta E \ll E$ is a small fixed energy that later will tend to zero, and choose in $\mathcal{E}_{\delta E}$ the constant density.

  (b) In this way we do not obtain a good ensemble because this is not ergodic (since it is a collection of invariant sets). However we obtain an approximation of an ergodic ensemble, because the energy variation δE is very small, and the density (which is an integral of motion) is constant.

(c) To obtain a correct definition of an ensemble we must now collapse on the manifold Σ_E by a limiting procedure. For fixed values of E and V we define the *microcanonical partition function*:

$$\mathcal{Z}_{\mathrm{M}}(N, V, E) := \frac{1}{N!} \lim_{\delta E \to 0} \frac{\mathrm{Vol}(\mathcal{E}_{\delta E})}{\delta E}.$$

But then we have:

$$\mathcal{Z}_{\mathrm{M}}(N, V, E) = \frac{1}{N!} \lim_{\delta E \to 0} \frac{\delta E \, \mathrm{Area}(\Sigma_E)}{\delta E} = \frac{\mathrm{Area}(\Sigma_E)}{N!}.$$

▶ According to the general methodology of Gibbsian formalism we should now define a macroscopic thermodynamic potential in terms of the microcanonical partition function. Such potential, which must be extensive, is the *entropy*.

Definition 3.3

Let $(\mathcal{E}_{\mathrm{M}}, \rho_{\mathrm{M}})$ be the ME corresponding to the Hamiltonian (3.1).

1. *The* **entropy** *is defined by*

$$S(N, V, E) := \kappa \log \mathcal{Z}_{\mathrm{M}}(N, V, E), \tag{3.18}$$

where κ is the Boltzmann constant.

2. *The* **temperature** *is defined by*

$$T := \left(\frac{\partial S}{\partial E} \right)^{-1}. \tag{3.19}$$

3. *The* **pressure** *is defined by*

$$P := T \frac{\partial S}{\partial V}. \tag{3.20}$$

▶ Remarks:

- The logarithm makes the entropy an extensive quantity, as required by the thermodynamical formalism. Indeed, the partition function of a system obtained by the union of isoenergetic systems of identical particles is the product of the partition functions of the component systems.

- Our definition of T and P agrees with formula (1.4) which expresses the *first law of thermodynamics*:

$$-P \, \mathrm{d}V + T \, \mathrm{d}S = \mathrm{d}E.$$

Indeed

$$dS = \frac{\partial S}{\partial E}\, dE + \frac{\partial S}{\partial V}\, dV = \frac{1}{T}\, dE + \frac{P}{T}\, dV.$$

3.3.1 Fluctuations and the Maxwell distribution

▶ We now claim and prove a remarkable statement, highlighting the relation between the formalism of the ME and the equilibrium solutions of the Boltzmann transport equation. It is not too restrictive to consider only the case of the Maxwell distribution (2.13):

$$\varrho_0(p) = n \left(\frac{\beta}{2\,\pi\,m} \right)^{3/2} e^{-\beta\, \mathfrak{h}(p)}, \tag{3.21}$$

where

$$\beta := \frac{1}{\kappa\, T}, \qquad \mathfrak{h}(p) := \frac{\|p\|^2}{2\,m}.$$

Here the phase space is $\Delta := \Lambda \times \mathbb{R}^3$. We will prove in the language of the ME that, if $N \gg 1$, the fluctuations around the Maxwell distribution are small, and hence the probability that the system takes a state different from this particular distribution is very small.

▶ Recall that in the Boltzmann approach the phase space Δ is parametrized by canonical coordinates (q, p) (whose meaning is different from the canonical coordinates in Ω). Each particle has a representative point in Δ. At every time t the kinematic state of the gas is completely defined in terms of N points in Δ.

- Introduce a finite partition in cells $(\Delta_1, \ldots, \Delta_K)$, $K \in \mathbb{N}$, of Δ. We define $\mathrm{Vol}(\Delta_i) := z$ for all $i = 1, \ldots, K$, with $Kz = \mathrm{Vol}(\Delta)$. Note that z must be small w.r.t. $\mathrm{Vol}(\Delta)$, but sufficiently large that we can find in each cell representative points of a sufficiently large number of particles.

- Let n_i be the occupation number of Δ_i, namely the number of representative points in the cell Δ_i. Let n_i be subject to the conditions of the ME:

$$\sum_{i=1}^{K} n_i = N, \qquad \sum_{i=1}^{K} \varepsilon_i\, n_i = E, \qquad \varepsilon_i := \frac{\|p_i\|^2}{2\,m}, \tag{3.22}$$

  and $p_i \in \mathbb{R}^3$ is the momentum corresponding to the cell Δ_i.

- A K-tuple $n := (n_1, \ldots, n_K)$ defines a discrete distribution function

$$\varrho_0(p_i) := \frac{n_i}{z}. \tag{3.23}$$

  for each cell Δ_i, $i = 1, \ldots, K$.

▶ The following statement holds.

Theorem 3.5

1. *The distribution function (3.23) defines in the ME a bounded region whose volume is*

$$\mathcal{Z}_M(n) = \frac{N!}{n_1! \cdots n_K!} z^N. \tag{3.24}$$

   After an appropriate normalization $\mathcal{Z}_M(n)$ can be interpreted as a discretization of the probability distribution function of the ME.

2. *The Maxwell distribution (3.21) is the continuous limit of the discrete distribution obtained by maximizing the volume $\mathcal{Z}_M(n)$. It is therefore the most probable distribution.*

Proof. We prove both claims.

1. To each prescribed distribution of the N points in the cells $\Delta_1, \ldots, \Delta_K$ (hence to each microscopic state) there corresponds exactly one specific cell of volume z^N in the space Ω. It is then clear that $\mathcal{Z}_M(n)$ is proportional to z^N. Now, observe that to a K-tuple $n := (n_1, \ldots, n_K)$ there corresponds more than one microscopic state, and hence a larger volume in the space Ω. For example, interchanging two particles with representative points in two distinct cells, the numbers n_i do not change, but the representative point in the space Ω does. Indeed, nothing changes if we permute particles inside the same cell. Since $N!$ is the total number of permutations and $n_i!$ are those inside the cell z_i, which do not change the position of the representative volume element in the space Ω, we find that the total volume in Ω corresponding to a prescribed sequence of numbers $(n_1, \ldots, n_K)$ is (3.24).

2. We seek a sequence $(\bar{n}_1, \ldots, \bar{n}_K)$ maximizing $\mathcal{Z}_M(n)$ and therefore expressing the most probable macroscopic state w.r.t. the microcanonical distribution. Recall that $n_i \gg 1$. Using the Stirling approximation (2.11) and (3.24), we obtain

$$\log \mathcal{Z}_M(n) \approx -\sum_{i=1}^{K} n_i \log n_i + \text{const.} \tag{3.25}$$

Considering now the variables n_i as continuous variables, we seek the maximum of the function (3.25) taking into account the constraints expressed by (3.22). Hence we want to compute the extremal points of the function

$$F_{\lambda_1, \lambda_2}(n) := -\sum_{i=1}^{K} (n_i \log n_i - \lambda_1 n_i - \lambda_2 \varepsilon_i n_i),$$

where $\lambda_1, \lambda_2 \in \mathbb{R}$ are Lagrangian multipliers. Computation yields

$$\log \bar{n}_i + 1 + \lambda_1 + \varepsilon_i \lambda_2 = 0, \qquad i = 1, \ldots, K. \tag{3.26}$$

Note that

$$\frac{\partial^2 F_{\lambda_1, \lambda_2}}{\partial n_i \partial n_j} = -\frac{\delta_{ij}}{n_i} < 0,$$

and therefore the extremum of F_{λ_1, λ_2} is a maximum. Redefining the parameters λ_1, λ_2 we can write the solutions of (3.26) in the form

$$\overline{n}_i = \alpha \, e^{-\beta \, \varepsilon_i}, \qquad i = 1, \ldots, K, \tag{3.27}$$

where $\alpha, \beta > 0$ are two constants. The continuous limit of the discrete distribution (3.27) is precisely (3.21). This automatically leads to the determination of the constants α, β.

The Theorem is proved.　　　　　　　　　　　　　　　　　　　　　　　　■

3.3.2 Thermodynamics of a free ideal gas

▶ We now consider a free ideal gas described in the ME. The aim is to derive its thermodynamics, which must agree with the free ideal gas law $PV = N\kappa T$.

▶ First of all we recall a classical result of Analysis. Let S_r^{n-1} be a hypersphere of radius $r > 0$ embedded in $\mathbb{R}^n$. Then the volume and surface area of S_r^n are given respectively by

$$\mathrm{Vol}\left(S_r^{n-1}\right) = \frac{\pi^{n/2} \, r^n}{\Gamma(1 + n/2)}, \qquad \mathrm{Area}\left(S_r^{n-1}\right) = \frac{2 \, \pi^{n/2} \, r^{n-1}}{\Gamma(n/2)}. \tag{3.28}$$

▶ Consider a free ideal gas in the ME.

- The Hamiltonian is

$$\mathcal{H}(x) := \mathcal{H}_0(p_1, \ldots, p_N) + \varepsilon \, \mathcal{H}_{\mathrm{per}}(q_1, \ldots, q_N), \qquad |\varepsilon| \ll 1, \tag{3.29}$$

where

$$\mathcal{H}_0(p_1, \ldots, p_N) := \sum_{i=1}^{N} \frac{\|p_i\|^2}{2 \, m},$$

and the perturbation $\mathcal{H}_{\mathrm{per}}$, which is responsible for the ergodicity of the system, is neglected in the derivation of the thermodynamics.

- The ME ensemble has been defined in Definition 3.2. In particular the energy level set is

$$\Sigma_E := \left\{ x \in \Omega : \sum_{i=1}^{N} \|p_i\|^2 = 2 \, m \, E \right\},$$

with $E \geqslant 0$, and

$$\mathcal{Z}_{\mathrm{M}}(N,V,E) := \frac{1}{N!}\mathrm{Area}(\Sigma_E) = \frac{1}{N!}\int_{\Sigma_E}\frac{\mathrm{d}\sigma}{\left\|\mathrm{grad}_{(p_1,\ldots,p_N)}\mathcal{H}_0(p_1,\ldots,p_N)\right\|},$$

where $\mathrm{d}\sigma$ is an infinitesimal element of the energy surface Σ_E.

(a) Note that

$$\mathrm{grad}_{(p_1,\ldots,p_N)}\mathcal{H}_0(p_1,\ldots,p_N) = \frac{1}{m}(p_1,\ldots,p_N),$$

so that

$$\left\|\mathrm{grad}_{(p_1,\ldots,p_N)}\mathcal{H}_0(p_1,\ldots,p_N)\right\|^2 = \frac{1}{m^2}\sum_{i=1}^{N}\|p_i\|^2 = \frac{2E}{m}.$$

(b) Therefore we get

$$\mathcal{Z}_{\mathrm{M}}(N,V,E) := \frac{V^N}{N!}\left(\frac{m}{2E}\right)^{1/2}\int_{\Sigma_E}\mathrm{d}\tilde{\sigma},$$

where we factorized the trivial integration in $(q_1,\ldots,q_N)$, giving the factor V^N, and $\mathrm{d}\tilde{\sigma}$ now denotes an infinitesimal element of Σ_E only in the momentum coordinates. By using the second formula (3.28) we find

$$\int_{\Sigma_E}\mathrm{d}\tilde{\sigma} = \frac{2\,\pi^{3N/2}(2\,m\,E)^{(3N-1)/2}}{\Gamma(3\,N/2)},$$

so that

$$\mathcal{Z}_{\mathrm{M}}(N,V,E) = \frac{V^N}{N!}\frac{1}{E}\frac{(2\,\pi\,m\,E)^{3N/2}}{\Gamma(3\,N/2)}. \tag{3.30}$$

▶ The next claim, known as Sackur-Tetrode formula, gives the entropy of the free ideal gas in terms of the microcanonical partition function (3.30).

Theorem 3.6 (*Sackur-Tetrode*)

For $N \gg 1$ the entropy of a free ideal gas is

$$S(N,V,E) \approx \kappa\,N\log\left(\frac{V}{N}\left(\frac{4\,\pi\,m\,E}{3\,N}\right)^{3/2}\right) + \frac{5}{2}\,\kappa\,N. \tag{3.31}$$

Proof. Formula (3.18) gives the exact formula

$$S(N, V, E) := \kappa \log \left(\frac{V^N}{N!} \frac{1}{E} \frac{(2 \pi m E)^{3N/2}}{\Gamma(3 N/2)} \right). \tag{3.32}$$

Recalling that $N \gg 1$ we use Stirling approximations (2.10) and (2.11) to estimate the quantity in brackets in formula (3.32). A bit of algebra gives

$$\frac{V^N}{N!} \frac{1}{E} \frac{(2 \pi m E)^{3N/2}}{\Gamma(3 N/2)} \approx \frac{V^N (2 \pi m E)^{3N/2}}{E} \frac{(2 \pi N)^{-1/2} N^{-N} e^N}{(2 \pi)^{1/2} (3 N/2)^{3N/2} e^{-3N/2}}$$

$$= \frac{1}{\sqrt{6} \pi E N} \left(\frac{V}{N} e^{5/2} \left(\frac{4 \pi m E}{3 N} \right)^{3/2} \right)^N.$$

Therefore, if $N \gg 1$, formula (3.32) gives the desired expression. ∎

▶ Remarks:

- The factor $1/N!$ (*Boltzmann counting*) in the partition function allows one to get an extensive entropy. Gibbs was the first to compute the entropy of the free ideal gas by using his theory. His result was the following formula:

$$S(N, V, E) \approx \kappa N \log \left(V \left(\frac{4 \pi m E}{3 N} \right)^{3/2} \right) + \frac{3}{2} \kappa N, \tag{3.33}$$

which coincides with (3.31) up to a replacement $V \mapsto V/N$ and an additional term κN. The apparently innocent replacement $V \mapsto V/N$ makes a crucial difference. Indeed the entropy (3.33) is not extensive and hence unacceptable as a thermodynamic potential of the system. If we consider two systems with the same particle density $n := N_1/V_1 = N_2/V_2$ and the same average energy per particle $\varepsilon := E_1/N_1 = E_2/N_2$, we want the entropy of their union to be the sum of the entropies S_1 and S_2. The entropy (3.33) does not have this property, and yields the paradoxical consequence that it is not possible to partition the system into two or more parts with identical ratios E_i/N_i, N_i/V_i and then re-assemble it, again obtaining the starting entropy (*Gibbs paradox*). This difficulty was immediately evident to Gibbs himself and he had no choice but to correct (3.33) by inserting V/N in place of V.

- The entropy (3.31) allows us to obtain the correct termodynamics. In the ME we can use formulas (3.19) and (3.20) to obtain

$$T := \left(\frac{\partial S}{\partial E} \right)^{-1} = \frac{2 E}{3 \kappa N} \qquad P := T \frac{\partial S}{\partial V} = \frac{2 E}{3 V},$$

which give the *free ideal gas law*

$$PV = N\kappa T,$$

and (cf. (2.22))

$$E = \frac{3}{2} N \kappa T. \tag{3.34}$$

- Theorem 3.6 and its consequences on the thermodynamics are valid *only* under the condition $N \gg 1$. Such limiting condition, which is indeed the TL in the ME, guarantees the orthodicity of the ME.

Example 3.5 (*An alternative definition of entropy*)

Consider a free ideal gas.

- Instead of using (3.18), which leads to

$$S(N, V, E) := \kappa \log \left(\frac{\mathrm{Area}(\Sigma_E)}{N!} \right),$$

we define

$$\widetilde{S}(N, V, E) := \kappa \log \left(\frac{\mathrm{Vol}(\Omega_E)}{N!} \right),$$

where $\Omega_E := \{x \in \Omega : 0 \leqslant \mathscr{H}(x) \leqslant E\}$ is the compact region enclosed by Σ_E.

- By standard integration we find

$$\mathrm{Vol}(\Omega_E) = V^N \frac{2 (2\pi m E)^{3N/2}}{3 N \Gamma(3N/2)} = \frac{2E}{3N} \mathrm{Area}(\Sigma_E).$$

- Then we have

$$\widetilde{S}(N, V, E) = \kappa \log \left(\frac{2E}{3N} \right) + S(N, V, E),$$

so that, if $N \gg 1$, the first contribution disappears.

3.4 Canonical ensemble

▶ Let us recall the physical situation corresponding to the canonical ensemble (CE):

- The system, whose (ergodic) Hamiltonian is (3.1), is closed, i.e., N is constant, but not isolated. It is maintained in thermal contact with a thermostat, a much larger system at fixed temperature T. Thermal contact means that the system can exchange energy through an interaction which must be weak as to not significantly perturb the microstates of the system.

- The system cannot exchange particles with its environment but can exchange energy with the thermostat, so that various possible states of the system can differ in total energy. In other words, there exist some small external random perturbations which make the internal energy of the system fluctuate around an average energy.

- The thermodynamic variables are the total number of particles N, the volume V and the temperature T (or the inverse temperature $\beta := (\kappa T)^{-1}$).

▶ We give the following definition.

Definition 3.4

> The **canonical ensemble** is defined by
>
> $$(\mathcal{E}_C, \rho_C) := \left(\Omega, \frac{e^{-\beta \mathscr{H}(x)}}{N!} \right). \tag{3.35}$$

▶ Remarks:

- As in the case of the ME, the probability distribution in (3.35) should be corrected with an overall factor which adjusts the dimensions. We will neglect this correction.

- The term $1/N!$ gives the correct counting of particles.

- Let us stress the fact that the Hamiltonian $\mathscr{H}$ is not constant along the trajectories in Ω, but instead fluctuates because of the interaction with a thermostat, which is perceived as a small random perturbation. In other words, we associate to $\mathscr{H}$ a statistical (non-deterministic) motion.

- The probability distribution function of the CE is

$$\widetilde{\rho}_C(x) := \frac{1}{\mathcal{Z}_C(N, V, \beta)} \frac{e^{-\beta \mathscr{H}(x)}}{N!},$$

where

$$\mathcal{Z}_C(N, V, \beta) := \frac{1}{N!} \int_{\Omega} e^{-\beta \mathscr{H}(x)} \, dx \tag{3.36}$$

is the *canonical partition function*. It is here assumed convergence of the integral.

- The CE-average of a measurable function $f : \Omega \to \mathbb{R}$ is by definition

$$\langle f(x) \rangle_C := \frac{1}{\mathcal{Z}_C(N, V, \beta)} \frac{1}{N!} \int_{\Omega} f(x) e^{-\beta \mathscr{H}(x)} \, dx.$$

- The definition of ρ_C (and of $\mathcal{Z}_C$) can be justified as follows. For simplicity we assume that there are no external fields acting on our system. In particular, such simplification will imply that our Hamiltonian depends only on momenta.

(a) At equilibrium the gas is described by the Maxwell distribution (2.13):

$$\varrho_0(p) = n \left(\frac{\beta}{2\pi m} \right)^{3/2} e^{-\beta \mathfrak{h}(p)}, \tag{3.37}$$

where

$$\beta := \frac{1}{\kappa T}, \qquad \mathfrak{h}(p) := \frac{\|p\|^2}{2m}.$$

Here the phase space is $\Delta := \Lambda \times \mathbb{R}^3$.

(b) Introduce a finite partition in cells $(\Delta_1, \ldots, \Delta_K)$, $K \in \mathbb{N}$, of Δ. We define $\mathrm{Vol}(\Delta_i) := z$ for all $i = 1, \ldots, K$, with $Kz = \mathrm{Vol}(\Delta)$. We know from Theorem 3.5 that to such a discretization there corresponds a discretization of the phase space Ω whose total volume is proportional to z^N.

(c) Let n_i, the occupation number of Δ_i, be subject to the conditions of the ME:

$$\sum_{i=1}^K n_i = N, \qquad \sum_{i=1}^K \mathfrak{h}_i \, n_i = E, \qquad \mathfrak{h}_i := \frac{\|p_i\|^2}{2m},$$

where $p_i \in \mathbb{R}^3$ is the momentum corresponding to the cell Δ_i.

(d) For a fixed cell $\Omega_0 \subset \Omega$, by projection on the component subspaces in Δ we can reconstruct the corresponding sequence $(n_1, \ldots, n_K)$ of occupation numbers in the cells $(\Delta_1, \ldots, \Delta_K)$. Such correspondence is not one to one, but we are only interested in using it to obtain information about the probability of finding a sampling of a representative point $x \in \Omega$ precisely in the cell Ω_0. This probability is the product of the probabilities of finding, in the space Δ, n_j points in the cell Δ_j for all $j = 1, \ldots, K$.

(e) According to (3.37), such a product is proportional to

$$\exp\left(-\beta \sum_{i=1}^K n_i \, \mathfrak{h}_i \right) = e^{-\beta \mathscr{H}},$$

where

$$\mathscr{H} := \sum_{i=1}^K n_i \, \mathfrak{h}_i$$

can be interpreted as a Hamiltonian in the space Ω. These heuristic arguments justify the form of the canonical density.

Example 3.6 (*The harmonic oscillator*)

Consider the Hamiltonian of a one-dimensional harmonic oscillator:

$$\mathscr{H}(q, p) := \frac{1}{2} \left(\frac{p^2}{m} + m\omega^2 q^2 \right).$$

Assume that the system is at temperature T. The phase space is $\Omega = \mathbb{R}^2$.

- The canonical partition function is given by

$$\mathcal{Z}_C(\beta) := \int_{\mathbb{R}^2} \exp\left(-\frac{\beta}{2}\left(\frac{p^2}{m} + m\,\omega^2\,q^2\right)\right) dq\,dp = \frac{2\,\pi}{\omega\,\beta}.$$

- We can arrive at the same result by using action-angle variables (see Example 3.3), so that the Hamiltonian is $\mathcal{H}(J) := \omega\,J$ and the phase space is $\Omega = \mathbb{R}_+ \times S^1$. Then we have

$$\mathcal{Z}_C(\beta) := \int_0^{2\pi} d\chi \int_0^{+\infty} e^{\beta\,\omega\,J}dJ = \frac{2\,\pi}{\omega\,\beta}.$$

▶ In the CE the total energy E is not fixed. It is rather a random variable distributed according to the distribution ρ_C. The next claim establishes the CE-average of the Hamiltonian, which is identified with E.

Theorem 3.7

Let $(\mathcal{E}_C, \rho_C)$ be the CE corresponding to the Hamiltonian (3.1). The CE-average of the energy is given by

$$E := \langle\,\mathcal{H}(x)\,\rangle_C = -\frac{\partial}{\partial\beta}\log\mathcal{Z}_C(N, V, \beta).$$

Proof. We have

$$\log\mathcal{Z}_C(N, V, \beta) := \log\left(\frac{1}{N!}\int_\Omega e^{-\beta\,\mathcal{H}(x)}\,dx\right),$$

so that

$$\frac{\partial}{\partial\beta}\log\mathcal{Z}_C(N, V, \beta) = -\frac{\displaystyle\int_\Omega \mathcal{H}(x)\,e^{-\beta\,\mathcal{H}(x)}\,dx}{\displaystyle\int_\Omega e^{-\beta\,\mathcal{H}(x)}\,dx}$$

$$=: -\langle\,\mathcal{H}(x)\,\rangle_C =: -E,$$

which is the claim.   ∎

▶ As we did for the ME we should now define a macroscopic (extensive) thermodynamic potential in terms of the canonical partition function. Such potential, in the case of the CE, is the *free energy* (or *Helmholtz energy*).

Definition 3.5

Let $(\mathcal{E}_C, \rho_C)$ be the CE corresponding to the Hamiltonian (3.1). The **free energy** is defined by

$$F(N, V, \beta) := -\frac{1}{\beta} \log \mathcal{Z}_C(N, V, \beta). \tag{3.38}$$

▶ The fact that F (or any thermodynamic potential) is extensive means that for fixed β and for large N and V such that the density $n := N/V$ is fixed, the TL defined by

$$\varphi(n, \beta) := -\frac{1}{\beta} \lim_{\substack{V, N \to +\infty \\ n \text{ fixed}}} \frac{1}{N} F(N, V, \beta) \tag{3.39}$$

exists and depends only on the intensive quantities n and β.

- The existence of such limit is not guaranteed a priori and it has to be checked case by case in order to assure the orthodicity of the ensemble. We see that existence of TL and extensive property of thermodynamic potentials are two sides of the same coin.

- We shall show that for suitable potential energies appearing in the Hamiltonian (3.1) the limit (3.39) exists. The extensive property of the entropy (see formula (3.41)) follows from the extensive property of the free energy.

▶ The next statement provides an alternative definition of the free energy.

Theorem 3.8

The free energy satisfies the following differential equation

$$F(N, V, \beta) - T \frac{\partial F}{\partial T} = E. \tag{3.40}$$

Proof. Formula (3.38) and Theorem 3.7 give:

$$E = -\frac{\partial}{\partial \beta} \left(-\beta F(N, V, \beta) \right) = F(N, V, \beta) - T \frac{\partial F}{\partial T},$$

where we used

$$\frac{\partial}{\partial \beta} = \frac{\partial T}{\partial \beta} \frac{\partial}{\partial T} = -\frac{1}{\kappa \beta^2} \frac{\partial}{\partial T} = -\kappa T^2 \frac{\partial}{\partial T}.$$

The claim is proved. ∎

▶ Remarks:

- A consistent definition of the entropy is

$$S(N, V, \beta) := -\frac{\partial F}{\partial T}. \tag{3.41}$$

- With the help of this definition one can write (3.40) as

$$F(N, V, \beta) = E - T\, S(N, V, \beta), \tag{3.42}$$

which can be seen as an alternative definition of the free energy.

- If E is independent of V, we can differentiate (3.42) w.r.t. V to get

$$-\frac{\partial F}{\partial V} = T\frac{\partial S}{\partial V} = P, \tag{3.43}$$

where we used formula (3.20). This formula provides a definition of the pressure in terms of the free energy.

- In this setting, there are at least other two important thermodynamic quantities: the *heat capacity at constant volume*,

$$C_V := \frac{\partial E}{\partial T} = -T\frac{\partial^2 F}{\partial T^2},$$

and the *isothermal compressibility*,

$$\chi_T := \left(-V\frac{\partial P}{\partial V}\right)^{-1} = \left(V\frac{\partial^2 F}{\partial V^2}\right)^{-1}. \tag{3.44}$$

Since E is an extensive quantity it follows that also C_V is extensive, i.e., it is proportional to N. It turns out that the conditions

$$0 < C_V < +\infty, \qquad 0 < \chi_T < +\infty, \tag{3.45}$$

(implying that F is a concave function of T and a convex function of V) are two necessary conditions for the *stability of thermodynamics*, which is, in fact, a consequence of the *second law of thermodynamics*. In this context the stability of thermodynamics means that the equilibrium state of the system is characterized by a stable minimum of F and two necessary stability conditions are exactly (3.45).

▶ The next Theorem indicates the equivalence of the ME and the CE under the TL $N \gg 1$.

Theorem 3.9

Let $(\mathcal{E}_C, \rho_C)$ be the CE corresponding to the Hamiltonian (3.1). Assume $0 < C_V < $

$+\infty$. *Under the TL $N \to +\infty$, the mean quadratic fluctuation of $\langle \mathscr{H}(x) \rangle_C$,*

$$\mathrm{mqf}(\mathscr{H}(x)) := \frac{\langle \mathscr{H}^2(x) \rangle_C - \langle \mathscr{H}(x) \rangle_C^2}{\langle \mathscr{H}(x) \rangle_C^2}, \tag{3.46}$$

goes to zero.

Proof. Recall that $E := \langle \mathscr{H}(x) \rangle_C$ is proportional to N. Then note that

$$
\begin{aligned}
-\frac{\partial}{\partial \beta} \langle \mathscr{H}(x) \rangle_C &= \frac{\partial^2}{\partial \beta^2} \log \mathcal{Z}_C(N,V,\beta) = \frac{\partial}{\partial \beta} \left(\frac{1}{\mathcal{Z}_C(N,V,\beta)} \frac{\partial \mathcal{Z}_C}{\partial \beta} \right) \\
&= -\frac{1}{\mathcal{Z}_C^2(N,V,\beta)} \left(\frac{\partial \mathcal{Z}_C}{\partial \beta} \right)^2 + \frac{1}{\mathcal{Z}_C(N,V,\beta)} \frac{\partial^2 \mathcal{Z}_C}{\partial \beta^2} \\
&= \left\langle \mathscr{H}^2(x) \right\rangle_C - \langle \mathscr{H}(x) \rangle_C^2.
\end{aligned}
$$

On the other hand,

$$-\frac{\partial}{\partial \beta} \langle \mathscr{H}(x) \rangle_C = \kappa T^2 \frac{\partial}{\partial T} \langle \mathscr{H}(x) \rangle_C =: \kappa T^2 C_V,$$

which is finite, positive and proportional to N. Therefore the difference $\langle \mathscr{H}^2(x) \rangle_C - \langle \mathscr{H}(x) \rangle_C^2$ is proportional to N and the mean quadratic fluctuation (3.46) goes to zero for $N \to +\infty$. ∎

▶ Theorem 3.9 says that the CE clusters around the ME under the TL $N \gg 1$. Therefore, if $N \gg 1$, we claim that:

- The states of the CE are concentrated in a "thin" region around the energy level set Σ_E, with $E := \langle \mathscr{H} \rangle_C$.

- The CE and ME are equivalent. They provide equivalent descriptions of the thermodynamics.

▶ The next statement is the so called "Theorem of the equipartition of the energy".

Theorem 3.10

Let $(\mathcal{E}_C, \rho_C)$ be the CE corresponding to the Hamiltonian (3.1). If $|\mathscr{H}(x)| \to +\infty$ as $\|x\| \to +\infty$ then the following formula holds true:

$$\left\langle x_i \frac{\partial \mathscr{H}}{\partial x_j} \right\rangle_C = \frac{\delta_{ij}}{\beta}, \qquad i,j = 1,\dots,2\ell N. \tag{3.47}$$

Proof. We have:

$$
\begin{aligned}
\left\langle x_i \frac{\partial \mathscr{H}}{\partial x_j} \right\rangle_{\mathrm{C}} &:= \frac{1}{\mathcal{Z}_{\mathrm{C}}(N,V,\beta)} \frac{1}{N!} \int_{\Omega} x_i \frac{\partial \mathscr{H}}{\partial x_j} \mathrm{e}^{-\beta \mathscr{H}(x)}\, \mathrm{d}x \\
&= -\frac{1}{\beta \mathcal{Z}_{\mathrm{C}}(N,V,\beta)} \frac{1}{N!} \int_{\Omega} x_i \frac{\partial}{\partial x_j} \left(\mathrm{e}^{-\beta \mathscr{H}(x)} \right) \mathrm{d}x \\
&= -\frac{1}{\beta \mathcal{Z}_{\mathrm{C}}(N,V,\beta)} \frac{1}{N!} \left(\left. x_i \mathrm{e}^{-\beta \mathscr{H}(x)} \right|_{\partial \Omega} - \delta_{ij} \int_{\Omega} \mathrm{e}^{-\beta \mathscr{H}(x)}\, \mathrm{d}x \right) \\
&= \frac{\delta_{ij}}{\beta},
\end{aligned}
$$

where we integrated by parts and used the boundary condition $|\mathscr{H}(x)| \to +\infty$ as $\|x\| \to +\infty$. ∎

Example 3.7 (*An application of Theorem 3.10*)

In the CE consider the Hamiltonian (perturbative potentials are here neglected)

$$
\mathscr{H}(x) := \frac{1}{2} \sum_{i=1}^{6N} a_i\, x_i^2,
$$

where $a_i \geq 0$. Let r be the number of non-zero coefficients a_i.

- Note that

$$
\sum_{i=1}^{6N} x_i \frac{\partial \mathscr{H}}{\partial x_i} = 2\, \mathscr{H}(x).
$$

- From Theorem 3.10 we obtain

$$
E := \langle \mathscr{H}(x) \rangle_{\mathrm{C}} = \frac{1}{2} \sum_{i=1}^{6N} \left\langle x_i \frac{\partial \mathscr{H}}{\partial x_i} \right\rangle_{\mathrm{C}} = \frac{r}{2\beta} = \frac{r}{2} \kappa\, T.
$$

Therefore each (non-zero) term in the Hamiltonian contributes to the average energy by the quantity $\kappa\, T/2$.

▶ It might be surprising that Theorem 3.10 is still valid for systems with a small number of particles (even one). In such cases one considers systems governed by a Hamiltonian whose trajectories in Ω do not follow the associated Hamiltonian flow, but a statistical motion subject to fluctuations.

Example 3.8 (*The harmonic oscillator*)

Consider the Hamiltonian of a one-dimensional harmonic oscillator:

$$
\mathscr{H}(q,p) := \frac{1}{2} \left(\frac{p^2}{m} + m\, \omega^2\, q^2 \right), \qquad m, \omega > 0.
$$

Assume that the system is at temperature T. The phase space is $\Omega = \mathbb{R}^2$.

- From the canonical partition function computed in Example 3.6 we find the CE-average of the energy:

$$E := \langle \mathscr{H}(x) \rangle_C = -\frac{\partial}{\partial \beta} \log \mathcal{Z}_C(\beta) = \frac{1}{\beta}.$$

The same result can be obtained from Theorem 3.10. One finds

$$\frac{1}{2m} \langle p^2 \rangle_C = \frac{1}{2\beta}, \qquad \frac{m\omega^2}{2} \langle q^2 \rangle_C = \frac{1}{2\beta},$$

which confirms the result.

- One easily finds

$$\mathrm{mqf}\left(\mathscr{H}(q,p)\right) := \frac{\langle \mathscr{H}^2(q,p) \rangle_C - \langle \mathscr{H}(q,p) \rangle_C^2}{\langle \mathscr{H}(q,p) \rangle_C^2} = 1.$$

This result is necessarily different from the analogous result for the deterministic motion, for which the energy is constant.

3.4.1 Thermodynamics of a free ideal gas

▶ We now consider a free ideal gas, described by the Hamiltonian (3.29), in the CE. We construct the canonical partition function and some thermodynamic quantities.

- From (3.36) we have

$$
\begin{aligned}
\mathcal{Z}_C(N,V,\beta) &= \frac{V^N}{N!} \left(\int_{\mathbb{R}^3} e^{-\beta \|\xi\|^2/(2m)} d\xi \right)^N \\
&= \frac{V^N}{N!} \left(4\pi \int_0^{+\infty} \|\xi\|^2 e^{-\beta \|\xi\|^2/(2m)} d\|\xi\| \right)^N \\
&= \frac{V^N}{N!} \left(2\pi \left(\frac{2m}{\beta}\right)^{3/2} \Gamma\left(\frac{3}{2}\right) \right)^N \\
&= \frac{V^N}{N!} \left(\frac{2\pi m}{\beta}\right)^{3N/2},
\end{aligned}
\tag{3.48}
$$

where we used factorization of integrals and standard integration in spherical coordinates (see also Lemma 2.1).

- By Theorem 3.7 the average energy is (cf. (2.22) and (3.34))

$$E := \langle \mathscr{H}(x) \rangle_C = -\frac{\partial}{\partial \beta} \log \mathcal{Z}_C(N,V,\beta) = \frac{3}{2} \frac{N}{\beta} = \frac{3}{2} N\kappa T, \tag{3.49}$$

which gives the heat capacity at constant volume:

$$C_V := \frac{\partial E}{\partial T} = \frac{3}{2} N\kappa.$$

- Formula (3.38) defines the free energy:

$$F(N, V, \beta) := -\frac{1}{\beta} \log \left(\frac{V^N}{N!} \left(\frac{2\pi m}{\beta} \right)^{3N/2} \right). \tag{3.50}$$

- Formula (3.41) gives the entropy

$$S(N, V, T) := \frac{\partial}{\partial T} \left(\kappa T \log \left(\frac{V^N}{N!} (2\pi m \kappa T)^{3N/2} \right) \right). \tag{3.51}$$

One can check the validity of formula (3.42).

- Defining the pressure as in (3.43) we get

$$P := -\frac{\partial F}{\partial V} = \frac{N}{\beta V},$$

which confirms the validity of the *free ideal gas law*.

▶ Remarks:

- Let us emphasize the fact that in the derivation of the above results we did not use the TL $N \gg 1$. One says that the CE is naturally orthodic.

- On the other hand, to check the CE and the ME provide the same thermodynamic behavior we impose the TL $N \gg 1$. Then the free energy (3.50) is

$$F(N, V, \beta) \approx -\frac{N}{\beta} \log \left(e \frac{V}{N} \left(\frac{2\pi m}{\beta} \right)^{3/2} \right).$$

Then the entropy (3.51) reduces to the Sackur-Tetrode formula (3.31):

$$\begin{aligned} S(N, V, E) &\approx \frac{\partial}{\partial T} \left(N \kappa T \log \left(e \frac{V}{N} (2\pi m \kappa T)^{3/2} \right) \right) \\ &= \kappa N \log \left(\frac{V}{N} \left(\frac{4\pi m E}{3N} \right)^{3/2} \right) + \frac{5}{2} \kappa N, \end{aligned}$$

where $E = (3 N \kappa T)/2$.

Example 3.9 (*The meaning of fluctuations*)

Consider a free ideal gas composed of a single particle, i.e., $N = 1$, with momentum $p_1 \in \mathbb{R}^3$. Then, Theorem 3.10 and formula (3.49) give

$$E := \frac{1}{2m} \left\langle \|p_1^2\| \right\rangle_{\mathrm{C}} = \frac{3}{2} \kappa T.$$

What is essentially different from the case of systems with many particles is the fact that the mean quadratic fluctuation of the energy is not small:

$$\frac{\langle \|p_1^2\|^2 \rangle_C - \langle \|p_1^2\| \rangle_C^2}{\langle \|p_1^2\| \rangle_C^2} = \frac{2}{3}.$$

On the other hand, for the total system of N particles one finds

$$\frac{\langle \mathcal{H}_0^2(p_1,\ldots,p_N) \rangle_C - \langle \mathcal{H}_0(p_1,\ldots,p_N) \rangle_C^2}{\langle \mathcal{H}_0(p_1,\ldots,p_N) \rangle_C^2} = \frac{2}{3N},$$

which is small if $N \gg 1$.

3.5 Grand canonical ensemble

▶ Let us recall the physical situation corresponding to the grand canonical ensemble (GE):

- The system, whose (ergodic) Hamiltonian is (3.1), is neither closed nor isolated and is maintained in thermodynamic equilibrium with a reservoir.

- The system can exchange energy and particles with the reservoir, so that various possible states of the system can differ in both their total energy and total number of particles.

- The thermodynamic variables are the chemical potential μ (or the *fugacity* $\zeta :=$ $e^{\beta\mu}$), the volume V and the temperature T (or the inverse temperature $\beta :=$ $(\kappa T)^{-1}$).

▶ The construction of the GE is done by using the CE and it turns out to be more involved than the costruction of the previous ensembles. Such complexity is mainly due to a more complex physical situation. Let us proceed by steps.

- Any open and non-isolated system (enclosed in a volume V at temperature T) can be idealized as a system obtained by eliminating a separation boundary between a closed system, labelled with 1, with N_1 particles and volume V_1 and a second much larger system (a thermostat), labelled with 2, with $N_2 \gg N_1$ particles and volume $V_2 \gg V_1$. When the separation boundary is removed the two systems can exchange particles and energy. In this setting the total energy energy and the number of particles N are random variables.

- We set $N := N_1 + N_2$ and $V := V_1 + V_2$. The temperature T is fixed by the thermostat. We are interested in finding the ensemble describing the system 1.

- The number of possible states of the total system coincides with the number of all possible decompositions of N and taking into account that, if N_1 is fixed, then there are $N!/(N_1! N_2!)$ permutations of the particles which give rise to distinct states in the two systems.

- Neglecting the interactions between particles in V_1 and particles in V_2, we can write the total Hamiltonian of the system as

$$\mathscr{H}_N(x) = \mathscr{H}_{N_1}(x_1) + \mathscr{H}_{N_2}(x_2),$$

where $x_i \in \Omega_{N_i}$, $i = 1, 2$, Ω_{N_i} being the phase space of the i-th system. Here $x \in \Omega_N := \Omega_{N_1} \cup \Omega_{N_2}$.

- If $\mathcal{Z}_1$, $\mathcal{Z}_2$ and $\mathcal{Z}$ (we omit the suffix C to simplify the notation) count the number of states in the systems, then the following relation must necessarily hold:

$$\mathcal{Z}(N, V, \beta) = \sum_{N_1=0}^{N} \mathcal{Z}_1(N_1, V_1, \beta) \, \mathcal{Z}_2(N_2, V_2, \beta),$$

that is

$$\mathcal{Z}(N, V, \beta) = \left(\frac{1}{N_2!} \int_{\Omega_{N_2}} e^{-\beta \mathscr{H}_{N_2}(x_2)} \, dx_2 \right) \sum_{N_1=0}^{N} \frac{1}{N_1!} \int_{\Omega_{N_1}} e^{-\beta \mathscr{H}_{N_1}(x_1)} \, dx_1.$$

- This suggests that we can take as density of the GE for system 1, the density of canonical ensemble with N_1 particles, corrected by a proper factor:

$$\rho_G(x_1, N_1, \beta) := \frac{\mathcal{Z}_2(N_2, V_2, \beta)}{\mathcal{Z}(N, V, \beta)} \frac{e^{-\beta \mathscr{H}_{N_1}(x_1)}}{N_1!}, \tag{3.52}$$

with

$$\int_{\Omega_{N_1}} \rho_G(x_1, N_1, \beta) \, dx_1 = \frac{\mathcal{Z}_1(N_1, V_1, \beta) \, \mathcal{Z}_2(N_2, V_2, \beta)}{\mathcal{Z}(N, V, \beta)}, \tag{3.53}$$

in such a way that ρ_G is normalized to one when summing over all possible states of system 1:

$$\sum_{N_1=0}^{N} \int_{\Omega_{N_1}} \rho_G(x_1, N_1, \beta) \, dx_1 = 1.$$

- We now write the correction factor in (3.52) in such a way that $\rho_G(x_1, N_1, \beta)$ depends only on the system 1.

  (a) By formula (3.38) we can write

$$\mathcal{Z}_2(N_2, V_2, \beta) = e^{-\beta F(N_2, V_2, \beta)} = e^{-\beta F(N-N_1, V-V_1, \beta)},$$
$$\mathcal{Z}(N, V, \beta) = e^{-\beta F(N, V, \beta)}.$$

  Thus

$$\frac{\mathcal{Z}_2(N_2, V_2, \beta)}{\mathcal{Z}(N, V, \beta)} = e^{-\beta \, (F(N-N_1, V-V_1, \beta) - F(N, V, \beta))}.$$

(b) Since $V_1 \ll V$ and $N_1 \ll N$ we can consider the first order expansion of the function $F(N - N_1, V - V_1, \beta) - F(N, V, \beta)$:

$$
\begin{aligned}
F(N - N_1, V - V_1, \beta) - F(N, V, \beta) &\approx -\frac{\partial F}{\partial N} N_1 - \frac{\partial F}{\partial V} V_1 \\
&=: -\mu N_1 + P V_1,
\end{aligned}
$$

where $\mu := \partial F / \partial N$ is called *chemical potential* and P is the pressure defined as in (3.43).

(c) Therefore

$$
\frac{\mathscr{Z}_2(N_2, V_2, \beta)}{\mathscr{Z}(N, V, \beta)} = e^{\beta \mu N_1 - \beta P V_1} = \zeta^{N_1} e^{-\beta P V_1}, \tag{3.54}
$$

where $\zeta := e^{\beta \mu}$ is called *fugacity*.

(d) Note that the pressure P is an intensive quantity, defined in the global set, but also in each of its parts, and it is therefore admissible to interpret it as the pressure of the system with N_1 particles. The same can be argued of the chemical potential. Hence (3.54) is expressed only through variables referring to the system 1 as desired:

$$
\rho_G(x_1, N_1, \beta) := \frac{\zeta^{N_1} e^{-\beta (\mathscr{H}_{N_1}(x_1) + P V_1)}}{N_1!}. \tag{3.55}
$$

▶ The above considerations allow us to drop the index 1 in (3.55) and to give a precise definition of the GE. In the same spirit of the definition of the CE, where the system is immersed in a heat bath, the GE is obtained by immersing a CE in a "particle bath", meaning that the particle number is no longer fixed. Let N be the number of particles in the CE at temperature T, volume V, pressure P, fugacity ζ, Ω_N be the phase space and $\mathscr{H}_N(x)$ be the Hamiltonian. Then we define the GE as follows.

Definition 3.6

The **grand canonical ensemble** *is defined by*

$$
(\mathcal{E}_G, \rho_G) := \left(\bigcup_{N=0}^{+\infty} \Omega_N, \frac{\zeta^N e^{-\beta (\mathscr{H}_N(x) + P V)}}{N!} \right). \tag{3.56}
$$

▶ Remarks:

- Note that it would be more correct to write $\rho_G|_{\Omega_N}$ instead of ρ_G. Indeed, the distribution function in (3.56) is the restriction of ρ_G to Ω_N.

- A reformulation of formula (3.53) is:

$$\int_{\Omega_N} \rho_G(x, N, \beta)\, dx = \zeta^N e^{-\beta PV} \mathcal{Z}_C(N, V, \beta), \tag{3.57}$$

which, summing over N, gives

$$1 = e^{-\beta PV} \sum_{N=0}^{+\infty} \zeta^N \mathcal{Z}_C(N, V, \beta). \tag{3.58}$$

- The probability of finding the system in any microscopic state with N particles is found by dividing the r.h.s. of (3.57) by the sum over N of the same expression. We therefore conclude that the probability that the number of particles of the system is N is given by

$$P(N) = \zeta^N \frac{\mathcal{Z}_C(N, V, \beta)}{\mathcal{Z}_G(\zeta, V, \beta)},$$

where the *grand canonical partition function* is defined by

$$\mathcal{Z}_G(\zeta, V, \beta) := \sum_{N=0}^{+\infty} \zeta^N \mathcal{Z}_C(N, V, \beta). \tag{3.59}$$

- The GE-average of a measurable function $f_N : \Omega_N \to \mathbb{R}$ is by definition

$$\langle f_N(x) \rangle_G := \frac{1}{\mathcal{Z}_G(\zeta, V, \beta)} \sum_{N=0}^{+\infty} \langle f_N(x) \rangle_C\, \zeta^N \mathcal{Z}_C(N, V, \beta).$$

- Formula (3.58) gives the so called *equation of state* of the system:

$$\beta PV = \log \mathcal{Z}_G(\zeta, V, \beta). \tag{3.60}$$

▶ In the GE the number of particles is not fixed. It is rather a random variable distributed according to the distribution ρ_G.

Theorem 3.11

Let $(\mathcal{E}_G, \rho_G)$ be the GE corresponding to the Hamiltonian (3.1). The GE-average of the number of particles is given by

$$\langle N \rangle_G = \zeta \frac{\partial}{\partial \zeta} \log \mathcal{Z}_G(\zeta, V, \beta). \tag{3.61}$$

Proof. Note that $\langle N \rangle_C = N$. By definition we have

$$\langle N \rangle_G := \frac{1}{\mathcal{Z}_G(\zeta, V, \beta)} \sum_{N=0}^{+\infty} N \zeta^N \mathcal{Z}_C(N, V, \beta).$$

On the other hand,

$$
\begin{aligned}
\zeta \frac{\partial}{\partial \zeta} \log \mathcal{Z}_G(\zeta, V, \beta) &= \frac{\zeta}{\mathcal{Z}_G(\zeta, V, \beta)} \sum_{N=0}^{+\infty} N \zeta^{N-1} \mathcal{Z}_C(N, V, \beta) \\
&= \frac{1}{\mathcal{Z}_G(\zeta, V, \beta)} \sum_{N=0}^{+\infty} N \zeta^N \mathcal{Z}_C(N, V, \beta), \quad (3.62)
\end{aligned}
$$

which proves the claim. ∎

▶ We now define a macroscopic (extensive) thermodynamic potential in terms of the grand canonical partition function.

Definition 3.7

Let $(\mathcal{E}_G, \rho_G)$ be the GE corresponding to the Hamiltonian (3.1). The **grand potential** is defined by

$$O(\zeta, V, \beta) := -\frac{1}{\beta} \log \mathcal{Z}_G(\zeta, V, \beta).$$

▶ To get a quantitative indication that the GE is equivalent to the CE we shall proceed as in Theorem 3.9, which says the fluctuations of $\langle \mathcal{H}(x) \rangle_C$ tend to zero under the TL $N \gg 1$ (if C_V is positive and finite). The equivalence between the CE and the GE can be argued in a similar way by stating that fluctuations of $\langle N \rangle_G$ go to zero under the TL.

- Let us identify $\langle N \rangle_G$ with $\overline{N}$, the most probable value of N. Note that $\overline{N}$ is determined by the dominant term in the series (3.59).

- Referring to definition (3.39) of TL we say that the system admits TL if for fixed density $\overline{n} := \overline{N}/V$ the limit

$$\varphi(\overline{n}, \beta) := -\frac{1}{\beta} \lim_{V \to +\infty} \frac{1}{\overline{n} V} \log \mathcal{Z}_C(\overline{n} V, V, \beta) \quad (3.63)$$

exists. Note that (3.63) defines the limiting value of the free energy of the system.

▶ The analog of Theorem 3.9 for the case of the GE is the next statement.

Theorem 3.12

Let $(\mathcal{E}_G, \rho_G)$ be the GE corresponding to the Hamiltonian (3.1). Assume $0 < \chi_T < +\infty$. If the TL (3.63) exists, then the mean quadratic fluctuation of $\langle N \rangle_G$,

$$\text{mqf}(N) := \frac{\langle N^2 \rangle_G - \langle N \rangle_G^2}{\langle N \rangle_G^2},$$

goes to zero for $N \to +\infty$.

Proof. We proceed by steps.

- From formula (3.62) we get

$$
\begin{aligned}
\zeta \frac{\partial}{\partial \zeta} \left(\zeta \frac{\partial}{\partial \zeta} \log \mathcal{Z}_G(\zeta, V, \beta) \right) &= \zeta \frac{\partial}{\partial \zeta} \frac{\displaystyle\sum_{N=0}^{+\infty} N \zeta^N \mathcal{Z}_C(N, V, \beta)}{\mathcal{Z}_G(\zeta, V, \beta)} \\
&= \frac{1}{\mathcal{Z}_G(\zeta, V, \beta)} \sum_{N=0}^{+\infty} N^2 \zeta^N \mathcal{Z}_C(N, V, \beta) \\
&\quad - \frac{1}{\mathcal{Z}_G^2(\zeta, V, \beta)} \left(\sum_{N=0}^{+\infty} N \zeta^N \mathcal{Z}_C(N, V, \beta) \right)^2 \\
&= \langle N^2 \rangle_G - \langle N \rangle_G^2. \qquad (3.64)
\end{aligned}
$$

- Recalling that $\beta \mu = \log \zeta$ we have

$$\zeta \frac{\partial}{\partial \zeta} = \zeta \frac{\partial \mu}{\partial \zeta} \frac{\partial}{\partial \mu} = \frac{1}{\beta} \frac{\partial}{\partial \mu}. \qquad (3.65)$$

- By using (3.60) and (3.65) we get

$$\zeta \frac{\partial}{\partial \zeta} \left(\zeta \frac{\partial}{\partial \zeta} \log \mathcal{Z}_G(\zeta, V, \beta) \right) = \beta V \zeta \frac{\partial}{\partial \zeta} \left(\zeta \frac{\partial P}{\partial \zeta} \right) = \frac{V}{\beta} \frac{\partial^2 P}{\partial \mu^2}. \qquad (3.66)$$

- Comparing (3.64) and (3.66) we get

$$\langle N^2 \rangle_G - \langle N \rangle_G^2 = \frac{V}{\beta} \frac{\partial^2 P}{\partial \mu^2}. \qquad (3.67)$$

- We need to express the r.h.s. of (3.67) in a more convenient way. Recall that

$$P := -\frac{\partial F}{\partial V}, \qquad \mu := \frac{\partial F}{\partial N}.$$

Assuming that the TL (3.63) exists we can write

$$F(V, N, \beta) = \overline{N}\,\varphi(v, \beta),$$

where $v = 1/\overline{n} = V/\overline{N}$. Note that

$$\frac{\partial}{\partial V} = \frac{1}{\overline{N}}\frac{\partial}{\partial v}, \qquad \frac{\partial}{\partial N} = \frac{\partial}{\partial \overline{N}}, \qquad \overline{N}\frac{\partial}{\partial \overline{N}} = -v\frac{\partial}{\partial v}.$$

Therefore, computation gives

$$P = -\frac{\partial \varphi}{\partial v}, \qquad \mu = \varphi(v, \beta) + v\,P,$$

from which we get

$$\frac{\partial \mu}{\partial v} = -v\frac{\partial \varphi}{\partial v^2} = v\frac{\partial P}{\partial v}, \qquad \frac{\partial P}{\partial \mu} = \frac{1}{v}, \qquad \frac{\partial^2 P}{\partial \mu^2} = \frac{\partial}{\partial v}\left(\frac{1}{v}\right)\frac{\partial v}{\partial \mu} = -\frac{1}{v^3}\left(\frac{\partial P}{\partial v}\right)^{-1}.$$

The last formula is related to the isothermal compressibility (3.44):

$$\frac{\partial^2 P}{\partial \mu^2} = \frac{\chi_T}{v^2}.$$

- Therefore we can write (3.67) as

$$\left\langle N^2 \right\rangle_{\mathrm{G}} - \left\langle N \right\rangle_{\mathrm{G}}^2 = \frac{\overline{N}\,\chi_T}{\beta\,v},$$

which is assumed to be finite and positive. Now the limit $N \to +\infty$ gives the claim.

The Theorem is proved.   ■

3.5.1 *Thermodynamics of a free ideal gas*

▶ We now consider a free ideal gas described in the GE.

- Formula (3.48) gives the canonical partition function:

$$\mathcal{Z}_{\mathrm{C}}(N, V, \beta) = \frac{V^N}{N!}\left(\frac{2\pi m}{\beta}\right)^{3N/2},$$

so that the grand canonical partition function is

$$\mathcal{Z}_{\mathrm{G}}(\zeta, V, \beta) = \sum_{N=0}^{+\infty}\frac{1}{N!}\left(\zeta V\left(\frac{2\pi m}{\beta}\right)^{3/2}\right)^N = \exp\left(\zeta V\left(\frac{2\pi m}{\beta}\right)^{3/2}\right).$$

- Formulas (3.60) and (3.61) give respectively

$$\beta P V = \zeta V \left(\frac{2 \pi m}{\beta} \right)^{3/2},$$

and

$$\langle N \rangle_{\mathrm{G}} = \zeta V \left(\frac{2 \pi m}{\beta} \right)^{3/2},$$

which combined together give the *free ideal gas law*.

3.6   Existence of the thermodynamic limit

▶ It is now obvious that for finite systems different ensembles produce different thermodynamic behaviors. So, from this point of view, the notion of TL is essential.

- The notion of infinite system is not at all trivial. The state of an infinite system is obtained as a result of a limiting procedure under which N and V tend to infinity (in some sense to be defined in a rigorous way) and the density $n :=$ N/V remains constant.

- It is also essential to understand for which 2-body interaction potential energies in the Hamiltonian (3.1) the TL does exist. We know that if all potential energies in (3.1) are seen as small perturbations, so that the gas is a free ideal gas, the TL is well defined and we recover the correct macroscopic thermodynamics. Nevertheless, it is natural to expect that if the potential energy describing the interaction between particles is not negligible (i.e., the gas is real), then the existence of the TL will depend on their analytic form and it can even not exist at all.

An exhaustive presentation of the theory of the TL lies outside the scope of this course. The presentation of the results which follow will give us just a flavor of the theory behind them.

▶ We start with some preliminary considerations on our Hamiltonian (3.1) (without external potentials):

$$\mathcal{H}(x) := \sum_{i=1}^{N} \frac{\|p_i\|^2}{2m} + \sum_{1 \leqslant i < j \leqslant N} \mathcal{U}(q_i - q_j), \tag{3.68}$$

where $\mathcal{U}$ is the 2-body interaction potential energy under investigation. In order to obtain from (3.68) an admissible thermodynamic behavior it is natural to impose the following conditions:

1. The interaction between distant particles must be negligible.

2. The interaction must not cause the collapse of infinitely many particles into a bounded region of Λ.

▶ The mathematical formulation of the above conditions is given in terms of *temperedness* and *stability* of $\mathscr{U}$.

Definition 3.8

Consider the Hamiltonian (3.68).

1. The interaction $\mathscr{U}$ is **tempered** if there exists $A \geqslant 0$, $R > 0$ and $s > 3$ such that

$$\mathscr{U}(q_i - q_j) \leqslant A \|q_i - q_j\|^{-s}, \tag{3.69}$$

for $\|q_i - q_j\| \geqslant R$ for all $i, j = 1, \ldots, N$.

2. The interaction $\mathscr{U}$ is **stable** if there exists $K \geqslant 0$ such that

$$\sum_{1 \leqslant i < j \leqslant N} \mathscr{U}(q_i - q_j) \geqslant -NK. \tag{3.70}$$

▶ Remarks:

- To get the idea of the meaning of temperedness we estimate with (3.69) the energy of interaction of a particle with other particles distributed randomly with constant density n at a distance $d \geqslant R$. Then

$$n \int_{\|y\| > d} \mathscr{U}(y)\, dy \leqslant A\, n \int_{\|y\| > d} \|y\|^{-s}\, dy = C \int_d^{+\infty} r^{2-s}\, dr,$$

where $C > 0$ is a constant. The r.h.s. of the above formula converges if $s > 3$ and goes to zero when $d \to +\infty$. Temperedness implies thus that the positive part of the interaction energy between particles at large distances is negligible. The negative part of the interaction energy is controlled by the stability condition (3.70).

- Consider a system governed by the Hamiltonian (3.68) in the GE and assume that $\mathscr{U}$ is stable. Then it is easy to see that the grand canonical partition function is convergent:

$$\mathcal{Z}_G(\zeta, V, \beta) = 1 + \sum_{N=1}^{+\infty} \frac{\zeta^N}{\lambda^{3N} N!} \int_{\Lambda^N} \exp\left(-\beta \sum_{1 \leqslant i < j \leqslant N} \mathscr{U}(q_i - q_j)\right) dq_1 \cdots dq_N$$

$$\leqslant 1 + \sum_{N=1}^{+\infty} \frac{\zeta^N}{\lambda^{3N} N!} V^N e^{\beta N K} = \exp\left(\frac{\zeta V e^{\beta K}}{\lambda^3}\right) < +\infty. \tag{3.71}$$

Here the factor $\lambda := (2\pi m/\beta)^{-1/2}$ comes from the kinetic part of the Hamiltonian and is called *thermal wavelength*. An interaction which violates the stability condition (3.70) is likely to lead to a non-thermodynamic behavior (*catastrophic interaction*). Indeed, the divergence of the partition function would mean that the probability of finding a certain number of particles inside a bounded region is zero.

Example 3.10 (*A catastrophic potential*)

Consider an interaction $\mathscr{U}$ which is central, i.e., $\mathscr{U}(q_i - q_j) = \mathscr{U}(\|q_i - q_j\|)$ for all $i, j = 1, \ldots, N$, and such that there exist constants $R > 0$ and $\mathscr{U}_0 > 0$ such that

$$
\begin{cases}
\mathscr{U}(\|q_i - q_j\|) = -\mathscr{U}_0 & \text{if } \|q_i - q_j\| < R, \\
\mathscr{U}(\|q_i - q_j\|) = 0 & \text{if } \|q_i - q_j\| \geqslant R,
\end{cases}
$$

for all $i, j = 1, \ldots, N$. Then,

$$
\sum_{1 \leqslant i < j \leqslant N} \mathscr{U}(q_i - q_j) = -\frac{N(N-1)}{2}\mathscr{U}_0,
$$

which violates the stability condition (3.70).

3.6.1 Van Hove interactions

▶ We now introduce some special 2-body interactions $\mathscr{U}$ which are both tempered and stable. We will see that systems with such interactions admit TL and their thermodynamics is stable. In particular, both the heat capacity at constant volume C_V and the isothermal compressibility χ_T exist and are positive.

Definition 3.9

Consider the Hamiltonian (3.68). The interaction $\mathscr{U}$ is a **Van Hove interaction** if it is central, i.e., $\mathscr{U}(q_i - q_j) = \mathscr{U}(\|q_i - q_j\|)$ for all $i, j = 1, \ldots, N$, and there exist constants $R_0 > 0$, $R_1 > R_0$ and $\mathscr{U}_0 > 0$ such that

$$
\begin{cases}
\mathscr{U}(\|q_i - q_j\|) = +\infty & \text{if } \|q_i - q_j\| \leqslant R_0, \\
-\mathscr{U}_0 \leqslant \mathscr{U}(\|q_i - q_j\|) < 0 & \text{if } R_0 < \|q_i - q_j\| < R_1, \\
\mathscr{U}(\|q_i - q_j\|) = 0 & \text{if } \|q_i - q_j\| \geqslant R_1,
\end{cases} \tag{3.72}
$$

for all $i, j = 1, \ldots, N$.

▶ Remarks:

- A Van Hove interaction is obviously tempered.

- A Van Hove interaction is stable. Indeed, only a finite number of particles, say $M < N$, can interact with a given particle. An upper bound for M is given by the number of spheres of radius R_0 which can be packed inside a sphere of radius R_1 (i.e., approximately R_1^3 / R_0^3). Since $\mathscr{U}(\|q_i - q_j\|) \geqslant -\mathscr{U}_0$ we have

$$\sum_{1 \leqslant i < j \leqslant N} \mathscr{U}(\|q_i - q_j\|) > -NM\mathscr{U}_0,$$

that is the condition defining a stable potential (with $K := M\mathscr{U}_0$).

▶ We want now to prove that if our Hamiltonian (3.68) is such that $\mathscr{U}$ is a Van Hove interaction then the system admits TL. To do so we first give a proper formulation of the problem.

- Consider a sequence of bounded regions $(\Lambda_\ell)_{\ell \geqslant 1}$, $\Lambda_\ell \subset \mathbb{R}^3$, $\text{Vol}(\Lambda_\ell) = V_\ell$, with $V_\ell < V_{\ell+1}$. Each region Λ_ℓ contains N_ℓ particles such that $n := N_\ell / V_\ell$ is fixed for all $\ell \geqslant 1$. We also define a constant specific volume $v := 1/n = V_\ell / N_\ell$.

- To each region Λ_ℓ we assign a local Hamiltonian

$$\mathscr{H}_\ell(x) := \sum_{i=1}^{N_\ell} \frac{\|p_i\|^2}{2m} + \sum_{1 \leqslant i < j \leqslant N_\ell} \mathscr{U}(q_i - q_j), \tag{3.73}$$

which leads to a local canonical partition function

$$\mathscr{Z}_{\mathrm{C}}(N_\ell, V_\ell, \beta) := \frac{1}{\lambda^{3N_\ell} N_\ell!} \int_{\Lambda_\ell^{N_\ell}} \exp\left(-\beta \sum_{1 \leqslant i < j \leqslant N_\ell} \mathscr{U}(q_i - q_j)\right) dq_1 \cdots dq_{N_\ell},$$

where $\lambda := (2\pi m/\beta)^{-1/2}$. To simplify the notation we set $\mathscr{Z}_\ell \equiv \mathscr{Z}_{\mathrm{C}}(N_\ell, V_\ell, \beta)$.

- According with the notion of TL (3.39) we define the *local free energy per particle* in the ℓ-th region,

$$\varphi_\ell(n, \beta) := -\frac{1}{\beta} \frac{1}{N_\ell} \log \mathscr{Z}_\ell. \tag{3.74}$$

- The problem is to provide a characterization of the regions $(\Lambda_\ell)_{\ell \geqslant 1}$ and of the interaction $\mathscr{U}$ in such a way that the limit defined by

$$\varphi(n, \beta) := \lim_{\ell \to +\infty} \varphi_\ell(n, \beta) \tag{3.75}$$

does exist. The quantity φ defines the *free energy per particle* of the total system.

▶ We have the following statement.

Theorem 3.13

> Consider a system governed by the Hamiltonian (3.68) where $\mathscr{U}$ is a Van Hove interaction. Then the system admits TL, i.e., the limit (3.75) exists.

Proof. We proceed by steps.

- We construct a special sequence of domains $(\Lambda_\ell)_{\ell \geqslant 1}$. Consider a cubic box $\Lambda_1 \subset \Lambda$ with "free" volume $V_1 := L^3$ and walls of thickness $R_0/2 < L/2$. It contains N_1 particles. Proceeding inductively we construct a larger cube $\Lambda_{\ell+1}$ by placing eight cubes Λ_ℓ with volumes V_ℓ and walls of thickness $R_0/2$. The "free" volume of $\Lambda_{\ell+1}$ is $V_{\ell+1} = 8 V_\ell$ and it contains $N_{\ell+1} = 8 N_\ell$ particles. By construction the density of particles in each domain is constant, i.e., $n := N_\ell/V_\ell$ is fixed for all $\ell \geqslant 1$.

- The canonical partition function in the domain Λ_ℓ is

$$\mathcal{Z}_\ell := \frac{1}{\lambda^{3N_\ell} N_\ell!} \int_{\Lambda_\ell^{N_\ell}} \exp\left(-\beta \sum_{1 \leqslant i < j \leqslant N_\ell} \mathscr{U}\left(\|q_i - q_j\|\right)\right) dq_1 \cdots dq_{N_\ell},$$

where $\lambda := (2\pi m/\beta)^{-1/2}$.

- The stability of $\mathscr{U}$ immediately implies

$$\mathcal{Z}_\ell < \frac{V_\ell^{N_\ell} e^{\beta N_\ell K}}{\lambda^{3N_\ell} N_\ell!}, \tag{3.76}$$

where $K > 0$ is the stability constant. In particular, condition (3.76) implies that if $N_\ell \gg 1$ one gets

$$\log \mathcal{Z}_\ell < N_\ell \log\left(\frac{V_\ell}{N_\ell} \frac{e^{\beta K + 1}}{\lambda^3}\right) = N_\ell \left(\beta K + 1 + \log\left(\frac{n}{\lambda^3}\right)\right), \tag{3.77}$$

where we used the approximation (2.11).

- Since $\mathscr{U}\left(\|q_i - q_j\|\right) \leqslant 0$ if q_i and q_j are in two distinct cubes we decrease the integrand in $\mathcal{Z}_{\ell+1}$ by eliminating interactions between particles in different cubes Λ_ℓ. The domain of integration is also decreased by restricting N_ℓ of the $N_{\ell+1}$ particles to be in each of the cubes Λ_ℓ. There are $(8 N_\ell)!/(N_\ell!)^8$ ways of arranging $8 N_\ell$ particles in eight cubes with N_ℓ in each. Therefore we get:

$$\mathcal{Z}_{\ell+1} > (\mathcal{Z}_\ell N_\ell!)^8 \frac{1}{(8 N_\ell)!} \frac{(8 N_\ell)!}{(N_\ell!)^8} = \mathcal{Z}_\ell^8,$$

so that

$$\log \mathcal{Z}_{\ell+1} > 8 \log \mathcal{Z}_\ell. \tag{3.78}$$

- Formulas (3.74) and (3.78) give

$$\varphi_{\ell+1}(n,\beta) := -\frac{1}{\beta}\frac{1}{N_{\ell+1}}\log \mathcal{Z}_{\ell+1} < -\frac{1}{\beta}\frac{1}{N_\ell}\log \mathcal{Z}_\ell =: \varphi_\ell(n,\beta), \qquad (3.79)$$

  where we used $N_{\ell+1} = 8\,N_\ell$. Formula (3.79) says that the sequence of local free energies per particle $(\varphi_\ell)_{\ell\geq 1}$ is monotonic non-increasing.

- We conclude noticing that $(\varphi_\ell)_{\ell\geq 1}$ is bounded from below thanks to condition (3.77). Indeed, we have

$$\varphi_\ell(n,\beta) := -\frac{1}{\beta}\frac{1}{N_\ell}\log \mathcal{Z}_\ell > -\frac{1}{\beta}\left(\beta K + 1 + \log\left(\frac{n}{\lambda^3}\right)\right),$$

  where the r.h.s. is a constant. Therefore the limit (3.75) exists.

The Theorem is proved. ∎

▶ To show that a system exhibits a thermodynamic behavior it is not sufficient to prove the existence of the TL, which guarantees the extensivity property of thermodynamic potentials. One must also show that the resulting thermodynamics is stable. In particular both the heat capacity at constant volume C_V and the isothermal compressibility χ_T must exist and they have to be positive. As already mentioned, the stability of thermodynamics is a consequence of convexity properties of thermodynamic potentials.

▶ We claim the next Theorem. The proof is omitted.

Theorem 3.14

> *Consider a system governed by the Hamiltonian (3.68) where $\mathcal{U}$ is a Van Hove interaction. Then its thermodynamics is stable. In particular:*
>
> 1. *The free energy per particle is a bounded concave function of the temperature T, i.e., C_V exists and is positive.*
>
> 2. *The free energy per particle is a bounded convex function of the specific volume $v := 1/n$, i.e., χ_T exists and is positive.*

No Proof.

3.7 The virial expansion

▶ The virial expansion expresses the pressure of a many-particle system in equilibrium as a power series in the particle density. It provides a generalization of the free ideal gas law.

▶ In our investigation of the free ideal gas in all Gibbs ensembles we recovered the free ideal gas law, which can be written as

$$P = \frac{n}{\beta}.$$

In the above form this is an expression for the pressure. Our task is now to consider a gas for which interaction energies between particles are not negligible. We will derive an approximated equation of state of the gas.

- We work in the GE. Let N be the number of particles in the CE at temperature T, volume V, pressure P, fugacity ζ, Ω_N be the phase space and $\mathcal{H}_N(x)$ be the Hamiltonian:

$$\mathcal{H}_N(x) := \sum_{i=1}^{N} \frac{\|p_i\|^2}{2m} + \sum_{1 \leqslant i < j \leqslant N} \mathcal{U}(q_i - q_j), \tag{3.80}$$

  where $\mathcal{U}$ is the 2-body interaction potential energy which is assumed to admit TL.

- As N varies from 0 to $+\infty$ (in the GE) we can write the single Hamiltonians (3.80) as

$$\mathcal{H}_0(x) := 0,$$

$$\mathcal{H}_1(x) := \frac{\|p_1\|^2}{2m},$$

$$\mathcal{H}_2(x) := \sum_{i=1}^{2} \frac{\|p_i\|^2}{2m} + \mathcal{U}(q_1 - q_2),$$

$$\mathcal{H}_3(x) := \sum_{i=1}^{3} \frac{\|p_i\|^2}{2m} + \mathcal{U}(q_1 - q_2) + \mathcal{U}(q_1 - q_3) + \mathcal{U}(q_2 - q_3),$$

and so on.

▶ The next claim provides a series expansion in powers of the density n (called *virial expansion*) of the pressure of a real gas.

Theorem 3.15

Set

$$w(\xi) := e^{-\beta \mathcal{U}(\xi)} - 1, \qquad \xi \in \mathbb{R}^3.$$

The equation of state of a real gas with Hamiltonian (3.80) is given by

$$P = \frac{n}{\beta} \left(J_1(\beta) + J_2(\beta)\, n + J_3(\beta)\, n^2 + O(n^3) \right),$$

where

$$J_1(\beta) := 1,$$

$$J_2(\beta) := -\frac{1}{2\,V} \int_{\Lambda^2} w(q_1 - q_2) \, dq_1 \, dq_2,$$

$$J_3(\beta) := -\frac{1}{3\,V} \int_{\Lambda^3} w(q_1 - q_2) \, w(q_1 - q_3) \, w(q_2 - q_3) \, dq_1 \, dq_2 \, dq_3.$$

Proof. We proceed by steps.

- Let us write the grand canonical partition function as

$$\mathcal{Z}_G(\zeta, V, \beta) = 1 + \zeta\,\mathcal{Z}_C(1, V, \beta) + \zeta^2\,\mathcal{Z}_C(2, V, \beta) + \zeta^3\,\mathcal{Z}_C(3, V, \beta) + O(\zeta^4).$$

- Expanding the equation of state (3.60) in powers of ζ we have

$$P = \frac{1}{\beta\,V} \log \mathcal{Z}_G(\zeta, V, \beta) = \frac{1}{\beta\,V} \left(\zeta\,\mathcal{Z}_1 + \zeta^2\,\mathcal{Z}_2 + \zeta^3\,\mathcal{Z}_3 + O(\zeta^4) \right), \qquad (3.81)$$

  where

$$\mathcal{Z}_1 := \mathcal{Z}_C(1, V, \beta),$$

$$\mathcal{Z}_2 := \mathcal{Z}_C(2, V, \beta) - \frac{1}{2}\mathcal{Z}_C^2(1, V, \beta),$$

$$\mathcal{Z}_3 := \mathcal{Z}_C(3, V, \beta) - \frac{1}{3}\mathcal{Z}_C(1, V, \beta)\,\mathcal{Z}_C(2, V, \beta)$$
$$+ \frac{1}{3}\mathcal{Z}_C(1, V, \beta) \left(\mathcal{Z}_C^2(1, V, \beta) - 2\,\mathcal{Z}_C(2, V, \beta) \right),$$

  and so on.

- To get P as a function of β, V, N we need to find an expression for the fugacity ζ. From (3.61) we get

$$N = \zeta \frac{\partial}{\partial \zeta} \log \mathcal{Z}_G(\zeta, V, \beta) = \left(\zeta\,\mathcal{Z}_1 + 2\,\zeta^2\,\mathcal{Z}_2 + 3\,\zeta_3\,\mathcal{Z}_3 + O(\zeta^4) \right), \qquad (3.82)$$

  which allows us to make the Ansatz

$$\zeta = \frac{N}{\mathcal{Z}_1} + A_1\,N^2 + A_2\,N^3 + O(N^4), \qquad (3.83)$$

  which should express ζ as a function of β, V, N. Inserting this Ansatz into (3.82) we can determine A_1 and A_2 by comparing coefficients of powers of N. In general, this procedure is complicated but it is straightforward at least for the first coefficients. One finds

$$A_1 = -2\frac{\mathcal{Z}_2}{\mathcal{Z}_1^3}, \qquad A_2 = -3\frac{\mathcal{Z}_3}{\mathcal{Z}_1^4} + 8\frac{\mathcal{Z}_2^2}{\mathcal{Z}_1^5}.$$

- Substituting (3.83) with the determined coefficients A_1, A_2 into (3.81) we can express P as a power expansion of $n := N/V$:

$$P = \frac{n}{\beta}\left(J_1(\beta) + J_2(\beta)\,n + J_3(\beta)\,n^2 + O(n^3)\right),$$

where

$$J_1(\beta) := 1, \qquad J_2(\beta) := -V\frac{\mathcal{Z}_2}{\mathcal{Z}_1^2}, \qquad J_3(\beta) := -2\,V^2\left(\frac{\mathcal{Z}_3}{\mathcal{Z}_1^3} - 2\frac{\mathcal{Z}_2^2}{\mathcal{Z}_1^4}\right).$$

- It remains to compute explicitly $\mathcal{Z}_1, \mathcal{Z}_2, \mathcal{Z}_3$. Let us illustrate the explicit computation of the second virial coefficient J_2. We know that (see (3.48))

$$\mathcal{Z}_1 := \mathcal{Z}_C(1, V, \beta) = \frac{V}{\lambda^3},$$

where $\lambda := (2\pi m/\beta)^{-1/2}$. Now we have

$$\mathcal{Z}_C(2, V, \beta) := \frac{1}{2\lambda^6}\int_{\Lambda^2} e^{-\beta\mathscr{U}(q_1 - q_2)}\,dq_1\,dq_2,$$

which allows to construct $\mathcal{Z}_2$. Then we get

$$
\begin{aligned}
J_2(\beta) &= -V\frac{2\,\mathcal{Z}_C(2, V, \beta) - \mathcal{Z}_C^2(1, V, \beta)}{2\,\mathcal{Z}_C^2(1, V, \beta)} \\
&= -\frac{1}{2V}\int_{\Lambda^2}\left(e^{-\beta\mathscr{U}(q_1 - q_2)} - 1\right)dq_1\,dq_2 \\
&= -\frac{1}{2V}\int_{\Lambda^2} w(q_1 - q_2)\,dq_1\,dq_2,
\end{aligned}
$$

which is the desired result. A similar computation gives the third virial coefficient J_3.

The Theorem is proved.   ∎

3.8   The problem of phase transitions

▶ One of the most interesting problems of statistical mechanics concerns phase transitions.

- They are ubiquitous in the physical world: the boiling of a liquid, the melting of a solid, the spontaneous magnetization of a magnetic material, up to the more exotic examples in superfluidity, superconductivity, and quantum chromodynamics.

- In its broadest sense, a phase transition happens any time a physical quantity, such a heat capacity, depends in a non-analytic (or non-differentiable, or discontinuous) way on some control parameter. In the vicinity of a phase transition point a small change in some control parameter (like the pressure of the temperature) results in a dramatic change of certain physical properties.

- An additional characteristic common feature of all phenomena involving phase transitions is the generation (or destruction) in the macroscopic scale of ordered structures, starting from microscopic short-range interactions. Moreover, in the regions of the space of the parameters corresponding to critical phenomena (hence in a neighborhood of a critical point), different systems have a similar behavior even quantitatively (*universality of critical behavior*).

▶ The mathematical theory and the physical understanding of phase transitions constitutes one of the most interesting and hardest problems of mathematical physics. An exhaustive presentation of the theory of phase transitions lies outside the scope of this course. Our plan is the following: we here present some general mathematical ideas which give a characterization of phase transitions. Then, in the next Chapter, we will study the prototypical model exhibiting phase transitions, namely the *two-dimensional Ising model*.

▶ In our study of Gibbsian theory of ensembles we understood several important facts. Some of them are:

- The thermodynamic behavior of a theoretical gas of particles can be deduced by constructing proper Gibbs ensembles. Distinct ensembles correspond to different boundary conditions used to describe the system. For finite systems, different ensemble provide different descriptions. Nevertheless, under a proper TL, the resulting thermodynamics does not depend on the ensemble, i.e., on the boundary conditions.

- The equivalence of ensembles has been characterized in terms of smallness of fluctuations of some characteristic random variables under the TL. We also found that such equivalence holds true provided that some characteristic thermodynamic quantities, precisely C_V and χ_T, are positive and finite.

Loosely speaking, phase transitions are characterized by the existence of some singular points of thermodynamic potentials, leading to divergent thermodynamic quantities as C_V and χ_T. From the mathematical point of view, such divergences can be detected only under the TL.

▶ A more rigorous definition (in the GE) of phase transition follows (note that ζ is here interpreted as a complex variable).

Definition 3.10

Let $(\mathcal{E}_G, \rho_G)$ be the GE corresponding to the Hamiltonian (3.1). Any singular point of the TL of the grand potential,

$$\omega(\zeta, v, \beta) := \lim_{\substack{V, N \to +\infty \\ v := V/N \text{ fixed}}} \frac{1}{V} \log \mathcal{Z}_G(\zeta, V, \beta), \qquad (3.84)$$

occurring for positive v and β is called a **phase transition point**.

▶ Remarks:

- The lack of a factor $-1/\beta$ in the r.h.s. of (3.84) is only a matter of traditional notation.

- A definition similar to 3.10 can be given in the CE in terms of singular points of the TL of the free energy, see (3.39).

- Recall that assuming stability of the interaction potential we have convergence of the grand canonical partition function, (see (3.71))

$$\mathcal{Z}_G(\zeta, V, \beta) \leqslant \exp\left(\frac{\zeta V e^{\beta K}}{\lambda^3}\right),$$

where $\lambda := (2\pi m/\beta)^{-1/2}$ and K is the stability constant. Such result proves not only the mentioned convergence, but also that the sum defining the partition function is an entire function of ζ for every finite V. Moreover, as every $\mathcal{Z}_C(N, V, \beta)$ forming $\mathcal{Z}_G(\zeta, V, \beta)$ is positive definite, there can be no zeroes of $\mathcal{Z}_G(\zeta, V, \beta)$ on the positive real ζ-axis for every finite V. Our conclusion is that the system does not exhibit any phase transition for finite V.

- If the TL of the grand potential exists we have the following relations:

$$\beta P = \omega(\zeta, v, \beta),$$

which follows from the equation of state (3.60), and

$$\frac{1}{v} = \lim_{\substack{V, N \to +\infty \\ v := V/N \text{ fixed}}} \frac{1}{V} \zeta \frac{\partial}{\partial \zeta} \log \mathcal{Z}_G(\zeta, V, \beta), \qquad (3.85)$$

which follows from (3.61).

▶ The next claim is one of the cornerstones of exact results of statistical mechanics.

Theorem 3.16 (Lee-Yang)

Let $(\mathcal{E}_G, \rho_G)$ be the GE corresponding to the Hamiltonian (3.1). Assume that the *interaction potential is stable.*

1. *If the surface area of the boundary of Λ increases no faster than $V^{2/3}$ then $\omega(\zeta, v, \beta)$ exists and is a continuous and monotonically increasing function of ζ on the positive real ζ-axis.*

2. *Let D be an open set of the complex ζ-plane containing a portion of the positive real ζ-axis and no zeroes of the TL of $\mathcal{Z}_G(\zeta, V, \beta)$ for any given V and T. Then the limit (3.84) converges uniformly in any closed set of D and $\omega(\zeta, v, \beta)$ is analytic in D.*

No Proof.

▶ Remarks:

- A *thermodynamic phase* is defined by those values of ζ contained in any single region D of Theorem 3.16.

- Since in any region D the convergence of the limit (3.84) is uniform we can interchange the order of lim and $\zeta(\partial/\partial\zeta)$ in (3.85). Therefore, in any single phase we can write

$$\beta P = \omega(\zeta, v, \beta), \qquad \frac{1}{v} = \zeta \frac{\partial\omega}{\partial\zeta}.$$

- Let us illustrate two possible characteristic behaviors:

  (a) Suppose that the TL of $\mathcal{Z}_G(\zeta, V, \beta)$ does not have any zero on the entire positive real ζ-axis. Then we can choose D so that it includes the entire positive real ζ-axis. In such a case the system always exists in a single phase.

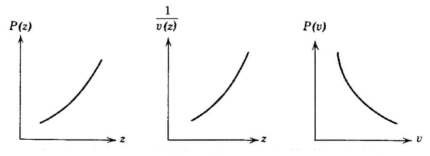

Fig. 3.2. Plots of $P(\zeta)$, $1/v(\zeta)$ and $P(v)$ in the case of a single phase ([Hu]).

(b) Let ζ_0 be a zero on the positive real ζ-axis of the TL of $\mathcal{Z}_G(\zeta, V, \beta)$. Then we can choose two distinct regions D_1 and D_2 in which the second claim of Theorem 3.16 holds separately.

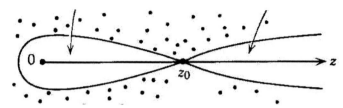

Fig. 3.3. A real zero ζ_0 of $\mathcal{Z}_G(\zeta, V, \beta)$ and two regions D_1, D_2 ([Hu]).

At $\zeta = \zeta_0$ the pressure P must be continuous, as required by the first claim of Theorem 3.16. However, the derivative w.r.t. ζ of P may be discontinuous. Therefore $1/v$ may be discontinuous. The system exhibits two phases, corresponding to the regions $\zeta < \zeta_0$ and $\zeta > \zeta_0$.

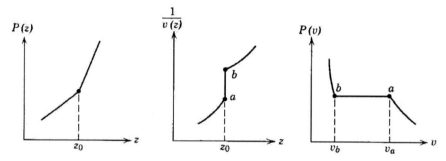

Fig. 3.4. Plots of $P(\zeta)$, $1/v(\zeta)$ and $P(v)$ in the case of a two phases ([Hu]).

3.9   Exercises

Ch3.E1 Consider a system composed by two distinct free ideal gases contained in two adjacent containers separated by a removable wall. The two gases have the same particle density $n := N_1/V_1 =
N_2/V_2$ and the same average energy per particle $\varepsilon := E_1/N_1 = E_2/N_2$. We remove the wall
separating the containers thus obtaining a mixture of the two gases. We want to compute the
change of entropy of the combined system in the formalism of the ME.

Recall that, in the case of a free ideal gas, the entropy is defined by

$$S(N, V, E) := \kappa \log \mathcal{Z}_M(N, V, E),$$

with

$$\mathcal{Z}_M(N, V, E) = \frac{V^N}{N!} \frac{1}{E} \frac{(2\pi m E)^{3N/2}}{\Gamma(3N/2)}. \tag{3.86}$$

We also consider a partition function $\widetilde{\mathcal{Z}}_M(N, V, E)$ which coincides with (3.86) up to Boltzmann's correction $1/N!$:

$$\widetilde{\mathcal{Z}}_M(N, V, E) = V^N \frac{1}{E} \frac{(2\pi m E)^{3N/2}}{\Gamma(3N/2)}. \tag{3.87}$$

Let S_i and S_f denote the entropies of the system before the wall is removed and after the wall
is removed.

(a) Compute the change of entropy $\Delta S := S_f - S_i$ by using both partition functions (3.86)
and (3.87).

(b) Perform the same computation in the case of two identical gases.

⁓⁓⁓⁓⁓⁓⁓⁓⁓⁓⁓⁓⁓⁓⁓

Ch3.E2 To appreciate the differences between the microcanonical and the canonical ensemble we consider the following problem. The final results obtained in the two ensembles must coincide.

A simple model for a polymer in two dimensions is a path on a lattice $\mathbb{Z}^2$. At every lattice point
the polymer can either go straight or choose between the two directions in a right angle with
respect to its current direction. Each time it bends in a right angle, it pays a bending energy
$\varepsilon > 0$. Thus, for a given shape (or configuration) of the polymer the total bending energy of the
polymer is ε times the number of right angle turns. We assume that the starting segment of the
polymer is fixed somewhere on $\mathbb{Z}^2$ and that the polymer consists of $N + 1$ segments, $N \gg 1$.
Each possible shape of the polymer is a state of this discrete system. Let T be the temperature
of the system.

• Microcanonical ensemble.

(a) Find the microcanonical partition function, namely the number $\mathcal{Z}_M(N, E)$ of polymer shapes that have a total bending energy $E := m\varepsilon, 0 \leqslant m \leqslant N, m \in \mathbb{N}$.

(b) Compute the entropy $S(N, E) := \kappa \log \mathcal{Z}_M(N, E)$ using Stirling's approximation.

(c) Calculate the temperature as a function of E and N.

(d) Express the energy E as a function of T and N.

• Canonical ensemble.

(a) Find the canonical partition function $\mathcal{Z}_C(N, T)$.

(b) Calculate the average internal energy as a function of T and N.

⁓⁓⁓⁓⁓⁓⁓⁓⁓⁓⁓⁓⁓⁓⁓

Ch3.E3 Consider a system of N identical but distinguishable particles, each of which has the two possible energy levels ε and $-\varepsilon$, $\varepsilon > 0$. Let T be the temperature of the system and assume that the system has fixed total energy E. Perform the following computations in the ME.

    (a) Compute the number of particles n_+ and n_- in the energy level ε and $-\varepsilon$ respectively in terms of N and E. We call n_+ and n_- occupation numbers.

    (b) Compute the entropy of the system using Stirling's approximation.

    (c) Explain how the previous two computations would change if the upper energy level had a g-fold degeneracy, while the lower energy level were non-degenerate.

    (d) Compute the free energy $F := E - TS$ as a function of T for the case of nondegenerate energy levels.

Ch3.E4 Consider a single particle of mass m at temperature T constrained to move on the surface of a sphere of radius r under the action of the gravitational field. The Hamiltonian governing the system is (in spherical coordinates)

$$\mathcal{H}(\vartheta, \varphi, p_\vartheta, p_\varphi) := \frac{1}{2\,m\,r^2} \left(p_\vartheta^2 + \frac{p_\varphi^2}{\sin^2 \vartheta} \right) + m\,g\,r\cos\vartheta, \qquad g > 0,$$

with $(\vartheta, \varphi, p_\vartheta, p_\varphi) \in [0, \pi) \times [0, 2\pi) \times \mathbb{R} \times \mathbb{R}$.

    (a) Compute the canonical partition function $\mathcal{Z}_C(r, \beta)$, where $\beta := (\kappa\,T)^{-1}$.

    (b) Compute the average energy $\langle \mathcal{H}(\vartheta, \varphi, p_\vartheta, p_\varphi) \rangle_C$.

    (c) Verify that at high temperatures the contribution of the gravitational potential energy is negligible.

Ch3.E5 Consider an ideal gas (N identical particles of mass m at temperature T) contained in a cubic box of side ℓ (resting on the horizontal plane $z = 0$) under the action of the gravitational field. The Hamiltonian governing the system is

$$\mathcal{H}(z_1, \ldots, z_N, p_1, \ldots, p_N) := \sum_{i=1}^{N} \frac{\|p_i\|^2}{2\,m} + m\,g \sum_{i=1}^{N} z_i,$$

where $p_i \in \mathbb{R}^3$ is the momentum of the i-th particle, $z_i \in [0, \ell]$ its z-coordinate and $g > 0$ the gravitational acceleration.

    (a) Compute the canonical partition function $\mathcal{Z}_C(N, V, \beta)$, where $\beta := (\kappa\,T)^{-1}$.

    (b) Compute the average energy $\langle \mathcal{H}(z_1, \ldots, z_N, p_1, \ldots, p_N) \rangle_C$.

    (c) Compute the average height $\langle z_i \rangle_C$.

    (d) Compute the free energy and the pressure.

    (e) Discuss the low and high temperature limits.

Ch3.E6 Consider a gas of N identical particles of mass m contained in a region of $\mathbb{R}^3$ with volume V at temperature T. Assume that the particles interact through a two-body central potential of the form

$$\mathscr{U}(|q_i - q_j|) := A|q_i - q_j|^{-\nu}, \qquad A > 0,\, \nu > 0, \qquad i, j = 1, \dots, N.$$

(a) Prove that the canonical partition function $\mathcal{Z}_C(N, V, T)$ satisfies the following functional equation:

$$\mathcal{Z}_C\left(N, \alpha^{-3/\nu} V, \alpha\, T\right) = \alpha^{3N(1/2 - 1/\nu)} \mathcal{Z}_C(N, V, T),$$

where $\alpha \in \mathbb{R} \setminus \{0\}$ is an arbitrary scaling factor.

(b) Prove that the free energy $F(N, V, T) := -\kappa\, T \log \mathcal{Z}_C(N, V, T)$ satisfies the following differential equation

$$T \frac{\partial F}{\partial T} - \frac{3}{\nu} V \frac{\partial F}{\partial V} = F - 3\left(\frac{1}{2} - \frac{1}{\nu}\right) N \kappa\, T.$$

(c) Prove that the internal energy E is related to the pressure P by the relation

$$E = x_1 P V + x_2 N \kappa\, T,$$

where x_1 and x_2 are functions of ν to be determined.

(d) Is there any limiting value of ν for which there holds $E = 3 N \kappa\, T / 2$? What is the corresponding value of A in the interaction $\mathscr{U}$ in order to make this limit (if exists) meaningful?

Ch3.E7 Consider a gas of N non-identical particles in $\mathbb{R}^d$, $d \geqslant 3$, with Hamiltonian

$$\mathscr{H}(q_1, \dots, q_N, p_1, \dots, p_N) := \sum_{i=1}^{N} \left(A_i \|p_i\|^s + B_i \|q_i\|^t\right),$$

with $A_i, B_i > 0$, $s, t \in \mathbb{N}$, $(q_i, p_i) \in \mathbb{R}^d \times \mathbb{R}^d$. Let T be the constant temperature of the system. Prove that

$$\langle \mathscr{H}(q_1, \dots, q_N, p_1, \dots, p_N) \rangle_C = N d \kappa\, T \left(\frac{1}{s} + \frac{1}{t}\right).$$

Ch3.E8 Consider a free ideal gas contained in a region of $\mathbb{R}^3$ of volume V at temperature T. The gas consists of two species, say 1 and 2, with $m_2 = 2 m_1 = 2 m$, and Hamiltonian per particle respectively given by

$$\mathscr{H}_1(p) := \frac{\|p\|^2}{2 m_1}, \qquad \mathscr{H}_2(p) := \frac{\|p\|^2}{2 m_2} + \delta,$$

where $\delta > 0$ is constant.

(a) Compute the grand canonical partition function.

(b) Compute the grand canonical potential.

(c) Compute the densities of particles of the two species.

Ch3.E9 A gas is in contact with a surface. On the surface there are N_0 localized and distinguishable sites adsorbing $N \leqslant N_0$ particles of the gas. Each site can adsorb zero or one particle of the gas. Let $\mathcal{Z}_C(\beta)$ be the canonical partition function of a single adsorbed particle and assume that all the adsorbed particles are non interacting.

(a) Compute the canonical partition function $\mathcal{Z}_C(N, \beta)$ of a system with N adsorbed particles.

(b) Compute the grand canonical partition function for all values of N from 0 to N_0:

$$\mathcal{Z}_G(\zeta, \beta) := \sum_{N=0}^{N_0} \zeta^N \mathcal{Z}_C(N, \beta),$$

where ζ is the fugacity.

(c) Compute the average number of particles, $\langle N \rangle_G$, adsorbed by the surface.

<hr />

Ch3.E10 Consider a system of N identical particles of mass m at temperature T ($\beta := (\kappa T)^{-1}$) moving on a straight line of length L and interacting through a two-body potential of the form

$$\mathcal{U}(|q_i - q_j|) := \begin{cases} \infty & |q_i - q_j| < a, \\ 0 & |q_i - q_j| > a, \end{cases}$$

with $0 < a < L$ and $i, j = 1, \ldots, N$. The configurational canonical partition function is

$$\mathcal{Z}_C(N, L, \beta) := \frac{1}{N!} \int_{[0,L]^N} \exp\left(-\beta \sum_{1 \leqslant i < j \leqslant N} \mathcal{U}(|q_i - q_j|)\right) dq_1 \cdots dq_N.$$

(a) Compute $\mathcal{Z}_C(N, L, \beta)$ explicitly.

(b) Compute the free energy per particle in the thermodynamic limit and prove that it is an analytic function of the density $v := L/N$ for $v > a$.

4

Introduction to Ising Models

4.1 Introduction

▶ There are innumerable solvable problems in classical mechanics, whereas, at the other extreme, very few solvable problems in statistical mechanics. The main reason for that is due to the fact that, in a system with a very large number of particles, each particle may directly or indirectly interact with an enormous number of others even if the fundamental interaction is 2-body and of short range.

▶ One of the most interesting aspects of statistical mechanics is the existence of phase transitions. Of the several existing models which exhibit phase transitions, the most famous one is probably the *Ising model*, a *classical spin system*.

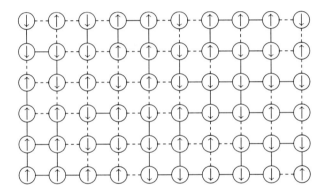

Fig. 4.1. A two-dimensional lattice with randomly oriented spins.

- The idea of a spin system was born around 1920 in an attempt to understand the phenomenon of *ferromagnetism*:

  (a) If we place a magnetic material (say an ordered lattice of iron atoms) in a magnetic field at fixed temperature, the field induces a certain amount of magnetization into the lattice, i.e., it creates a tendency for the *elementary magnetic momenta*, called *spins*, to point is a given direction. One can think of a spin as a discrete variable which takes values $+1$ and -1 or "up" ($\uparrow$) and "down" ($\downarrow$). The amount of magnetization depends on the strength of the field and on the temperature.

  (b) Now suppose that the external field is slowly turned off. For high temperatures, the lattice returns to an unmagnetized condition. But, for low temperatures, the lattice retains a degree of magnetism and the spins tend to preserve their coherent alignment. This phenomenon goes under the name of *spontaneous magnetization*.

(c) One can experimentally observe that there exists a *critical temperature* at which spontaneous magnetization begins to appear. This corresponds to a *phase transition* of the system, i.e., a transition between two different thermodynamic phases.

- It was understood that spins should exert an attractive (ferromagnetic) interaction among each others, which, however, is of short range. The question was then, how such a short range interaction could sustain the observed very long range coherent behavior of the material, and why such an effect should depend on the temperature.

- The Ising model was introduced by Lenz and Ising in 1925 to explain ferromagnetism. It is defined by a lattice configuration of a large number of spins, some boundary conditions and a configurational energy describing how spins interact among each other and with an external field. Of course, a physical lattice is three-dimensional, but also one- and two-dimensional lattices are admissible. For the mathematical and physical description of the Ising model the Gibbsian formalism of continuous systems does not apply. However, a discrete counterpart of it can be developed and applied.

  (a) In 1925 Ising succeeded in solving the one-dimensional model exactly and he found that there was no phase transition. This negative result gave (wrong) arguments in favor of non-existence of phase transitions in two and three dimensions.

  (b) In the 1930's Bragg, Williams, Bethe and Peierls and many others considered the two-dimensional Ising model as a model for binary alloys (spin up corresponding to an atom of type A and spin down to an atom of type B). In 1936 Peierls gave a proof of existence of ferromagnetism, but it was incorrect. Peierls' proof was corrected by Griffiths in 1964.

  (c) In 1942 Kramers and Wannier formulated the problem as a matrix problem and from symmetry considerations they were able to locate the phase transition point of a two-dimensional Ising model.

  (d) In 1944 Onsager solved completely the two-dimensional problem in the case of absence of external field. He used some algebraic techniques to compute the partition function and the free energy in the TL, thus proving in a rigorous way the existence of a phase transition. In particular, the TL of the free energy exhibits a logarithmic divergence at a given critical temperature. For such value of the temperature the heat capacity per spin diverges. Note that, as for continuous systems, no phase-transitions appear for finite systems.

  (e) After Onsager's works many other solutions for the two-dimensional case were proposed. In particular, beside the algebraic approach proposed by Onsager, a combinatorial approach gave the same results. The corresponding problem for the three-dimensional Ising model is unsolved, even in absence of external field.

- The Ising model has been applied to problems in chemistry, biology and other ares where "cooperative" behavior of large systems is studied. These applications are possible because the Ising system can be formulated as a general mathematical problem. Ferromagnetism is only one of its possible applications.

4.2 Definition of Ising models

▶ We start with the definition of two-dimensional Ising models.

- Let $\Lambda \subset \mathbb{Z}^2$ be a two-dimensional rectangular lattice with m rows and n columns. The lattice has $D := m\,n$ intersection points (*sites*).

- Λ can admit the following boundary conditions:

    1. *Free boundary conditions.* Λ is free in both horizontal and vertical directions.

    2. *Cylindrical boundary conditions.* Λ is free in one direction but cyclic in the other, i.e., Λ is wrapped on a cylinder.

    3. *Toroidal boundary conditions.* Λ is cyclic in both horizontal and vertical directions, i.e., Λ is wrapped on a torus.

- At each intersection site $\alpha \in \Lambda$ there is a *spin*, which is a discrete variable $\omega_\alpha = \pm 1$, ("up" ($\uparrow$) and "down" ($\downarrow$)). We parametrize α by $(i, j) \in \Lambda$, where i labels the rows and j labels the columns.

- There is a total of 2^D possible configurations of spins on Λ, a *configuration* $\{\omega\}$ being specified by the D spin variables, i.e.,

$$\{\omega\} := \{\omega_{i,j},\ (i,j) \in \Lambda\}.$$

Therefore the *phase space* of the system is

$$\Omega := \{+1, -1\}^D,$$

and $\{\omega\} \in \Omega$.

- The system is in thermal equilibrium at temperature $T > 0$.

- We assume that only *nearest neighbor spins* can interact among each other. In particular, any two nearest neighbor spins have a mutual constant interaction energy

$$-J_1\,\omega_{i,j}\,\omega_{i,j+1} = \begin{cases} -J_1 & \text{if } \omega_{i,j}\,\omega_{i,j+1} = +1, \\ +J_1 & \text{if } \omega_{i,j}\,\omega_{i,j+1} = -1, \end{cases}$$

and

$$-J_2\,\omega_{i,j}\,\omega_{i+1,j} = \begin{cases} -J_2 & \text{if } \omega_{i,j}\,\omega_{i+1,j} = +1, \\ +J_2 & \text{if } \omega_{i,j}\,\omega_{i+1,j} = -1, \end{cases}$$

with $(i,j) \in \Lambda$ and $J_1, J_2 \in \mathbb{R}$.

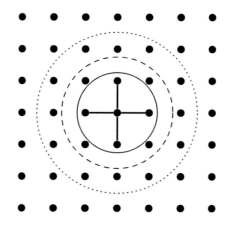

Fig. 4.2. A two-dimensional lattice and representation of nearest neighbor interactions.

- Each spin may interact with a constant external *magnetic field* with strength $H \in \mathbb{R}$. The corresponding interaction energy is

$$-H\,\omega_{i,j} = \begin{cases} -H & \text{if } \omega_{i,j} = +1, \\ +H & \text{if } \omega_{i,j} = -1. \end{cases}$$

- For a given configuration $\{\omega\} \in \Omega$ we define the *configurational energy* as the function

$$\mathscr{E}(\{\omega\}, J_1, J_2, H) := -\sum_{(i,j)\in\Lambda} \left(J_1\,\omega_{i,j}\,\omega_{i,j+1} + J_2\,\omega_{i,j}\,\omega_{i+1,j} + H\,\omega_{i,j} \right). \tag{4.1}$$

- Different spin configurations $\{\omega\} \in \Omega$ give different values of the configurational energy (4.1). Such different values are called *energy levels* and they span a energy spectrum, which is discrete.

▶ Remarks:

- The spin variables are time-independent. Hence $\mathscr{E}$ is not a Hamiltonian in the strict sense because it does not generate any Hamiltonian flow in Ω.

- The assumption of nearest neighbor interactions among spins is necessary for the solvability of the model. But it can be relaxed for the reduction to the one-dimensional case.

- The configurational energy has the following discrete symmetry:

$$\mathscr{E}(\{\omega\}, J_1, J_2, H) = (\{-\omega\}, J_1, J_2, -H). \tag{4.2}$$

- The *total spin* of the system is defined by:

$$\omega_{\text{tot}} := \sum_{(i,j) \in \Lambda} \omega_{i,j} = -\frac{\partial \mathscr{E}}{\partial H}.$$

- For the computation of the partition function we will assume $J_1 = J_2 =: J > 0$ (*ferromagnetic case*). In such a case, it is clear that we get a lower energy for a parallel configuration. In particular:

  (a) If toroidal boundary conditions are imposed on Λ, the *ground state energy*, i.e., the minimum of $\mathcal{E}$ is

  $$\mathscr{E}_0(J, H) := \min_{\{\omega\} \in \Omega} \mathscr{E}(\{\omega\}, J, H) = -D\left(2J + |H|\right).$$

  (b) If free boundary conditions are imposed on Λ, the ground state energy is

  $$\mathscr{E}_0(J, H) = -\left(D - \left\lfloor \frac{n+m}{2} \right\rfloor\right)\left(2J + |H|\right),$$

  where the boundary term which corrects D is negligible w.r.t. D for large m and n.

  We see that the minimum energy is achieved when all spins are up (down) if $H > 0$ (if $H < 0$). If $H = 0$ the minimum is achieved when all spins are either up or down. If $J_1 = J_2 =: J < 0$ (*antiferromagnetic case*) the ground state is not always easy to find.

4.3 Gibbsian formalism for Ising models

▶ Our task is to study the canonical distribution of the discrete energy levels corresponding to different spin configurations. Such distribution will be derived starting from the analysis of the model in the ME.

▶ We make the following assumptions on the discrete energy levels:

1. The discrete energy levels are commensurable.

2. All differences between energy levels are integral multiples of a quantity $\Delta > 0$.

3. Δ is the largest number satisfying 2.

If $H = 0$ the above conditions imply that $J_1 / J_2 \in \mathbb{Q}$.

▶ The construction of the ME is done in the same spirit illustrated in Chapters 2 (see Section 2.4) and 3. Our final result will be to extrapolate, in the TL, the canonical distribution of energy levels: the probability that the Ising model, at inverse temperature $\beta := (\kappa T)^{-1}$, has a spin configuration $\{\widetilde{\omega}\} \in \Omega$ is

$$P(\{\widetilde{\omega}\}, \beta, J_1, J_2, H) = \frac{e^{-\beta \mathscr{E}(\{\widetilde{\omega}\}, J_1, J_2, H)}}{\mathcal{Z}(\beta, J_1, J_2, H)},$$

where

$$\mathcal{Z}(\beta, J_1, J_2, H) := \sum_{\{\omega\} \in \Omega} e^{-\beta \mathscr{E}(\{\omega\}, J_1, J_2, H)}$$

is the *canonical partition function*. Here the sum is over all possible spin configurations on Ω. Hereafter, to simplify the notation, we will omit the arguments J_1, J_2, H from the list of dependencies of the configurational energy but we prefer to keep trace of them in the list of dependencies of the partition function.

- We introduce the concept of ensemble as a collection of $N \gg 1$ identical copies of the system, each of them with configuration $\{\omega\}_\ell \in \Omega$, $\ell = 1, \ldots, N$.

  (a) To each ℓ-th copy of the system we associate the configurational energy

  $$\mathscr{E}(\{\omega\}_\ell) := -\sum_{(i,j) \in \Lambda} \left(J_1\, \omega_{i,j}^{(\ell)}\, \omega_{i,j+1}^{(\ell)} + J_2\, \omega_{i,j}^{(\ell)}\, \omega_{i+1,j}^{(\ell)} + H\, \omega_{i,j}^{(\ell)} \right),$$

  where $\{\omega\}_\ell$ is free to vary on Ω.

  (b) The total configurational energy of the ensemble is defined by

  $$\mathscr{E}_{\text{tot}} := \sum_{\ell=1}^{N} \mathscr{E}(\{\omega\}_\ell), \tag{4.3}$$

  which gives a value $E_{\text{tot}} \in \mathbb{R}$, depending on J_1, J_2, H, if the ℓ configurations $\{\omega\}_\ell$ are fixed.

  (c) The average total configurational energy is defined by

  $$\overline{E} := \frac{E_{\text{tot}}}{N}.$$

- The working hypothesis is that all composing models are equiprobable. Therefore the probability that the N Ising models admit prescribed configurations $\{\omega\}_1, \ldots, \{\omega\}_N$ with total configurational energy $E_{\text{tot}} \in \mathbb{R}$ is:

$$P\left(\{\omega\}_1, \ldots, \{\omega\}_N, \beta, J_1, J_2, H, E_{\text{tot}}\right) := \frac{\delta_{\mathscr{E}_{\text{tot}}, E_{\text{tot}}}}{\mathcal{Z}_M},$$

where

$$\mathcal{Z}_M := \sum_{\{\omega\}_1 \in \Omega} \cdots \sum_{\{\omega\}_N \in \Omega} \delta_{\mathscr{E}_{\text{tot}}, E_{\text{tot}}}. \tag{4.4}$$

Here $\mathcal{Z}_M$ is the *microcanonical partition function*, δ is a Kronecker delta function and each local sum indicates a summation over all possible local configurations on Ω.

- We are not interested in the properties of all N Ising models, but rather in the properties of one particular member of the ensemble, say the first one ($\ell = 1$), which is seen as connected to an external "large" system of $N - 1$ Ising models (a "Ising model bath") at fixed inverse temperature β.

- The probability that the first Ising model has a configuration $\{\omega\}_1$ while the remaining $N - 1$ Ising model are in any equiprobable state subject only to the requirement that the total configurational energy (4.3) is constant, is given by

$$P\left(\{\omega\}_1, \beta, J_1, J_2, H, E_{\text{tot}}\right) := \frac{1}{\mathcal{Z}_M} \sum_{\{\omega\}_2 \in \Omega} \cdots \sum_{\{\omega\}_N \in \Omega} \delta_{\mathscr{E}_{\text{tot}}, E_{\text{tot}}}. \tag{4.5}$$

4.3.1 Canonical ensemble

▶ Our task is to get an approximate expression in the TL $N \gg 1$ for the exact formula (4.5). Such result will give us the desired canonical distribution of the energy levels.

Theorem 4.1

Consider the ME for Ising models defined above under the TL $N \gg 1$.

1. *The probability (4.5) is asymptotically given by*

$$P\left(\{\omega\}_1, \beta, J_1, J_2, H, E_{\text{tot}}\right) \approx \frac{e^{-\beta \mathscr{E}(\{\omega\}_1)}}{\mathcal{Z}(\beta, J_1, J_2, H)}, \tag{4.6}$$

where

$$\mathcal{Z}(\beta, J_1, J_2, H) := \sum_{\{\omega\} \in \Omega} e^{-\beta \mathscr{E}(\{\omega\})}. \tag{4.7}$$

The configuration $\{\omega\}$ appearing in (4.7) is any of the N spin configurations $\{\omega\}_\ell$ and the sum in (4.7) is over all possible spin configurations on Ω.

2. There holds

$$\overline{E} = -\frac{\partial}{\partial \beta} \log \mathcal{Z}(\beta, J_1, J_2, H). \tag{4.8}$$

Proof. The proof of the Theorem is based on rather technical asymptotic expansions. We will omit all details of such expansions and just give a sketch of the derivation.

• We use the following integral representation of the Kronecker delta:

$$\delta_{i,j} = \frac{1}{2\pi} \int_{-\pi}^{+\pi} e^{i\theta(i-j)} d\theta. \tag{4.9}$$

• Thanks to our assumptions on the discrete energy levels we can use (4.9) to write the microcanonical partition function (4.4) as

$$
\begin{aligned}
\mathcal{Z}_{\mathrm{M}} &= \frac{1}{2\pi} \sum_{\{\omega\}_1 \in \Omega} \cdots \sum_{\{\omega\}_N \in \Omega} \int_{-\pi}^{+\pi} \exp\left(\frac{i\theta(E_{\mathrm{tot}} - \mathcal{E}_{\mathrm{tot}})}{\Delta} \right) d\theta \\
&= \frac{\Delta}{2\pi i} \int_{-i\pi/\Delta}^{+i\pi/\Delta} e^{\eta E_{\mathrm{tot}}} \sum_{\{\omega\}_1 \in \Omega} \cdots \sum_{\{\omega\}_N \in \Omega} \exp\left(-\eta \sum_{\ell=1}^{N} \mathcal{E}(\{\omega\}_\ell) \right) d\eta \\
&= \frac{\Delta}{2\pi i} \int_{-i\pi/\Delta}^{+i\pi/\Delta} e^{\eta E_{\mathrm{tot}}} \prod_{j=1}^{N} \sum_{\{\omega\}_j \in \Omega} e^{-\eta \mathcal{E}(\{\omega\}_j)} d\eta.
\end{aligned}
$$

We used the change of variable defined by $\theta \mapsto \eta := i\theta/\Delta$.

• Since we are only interested in the first Ising model and not in the Ising model bath, we can describe the Ising bath of $N-1$ Ising models by any fiction we want provided that thermal equilibrium is maintained. In particular, we can assume that the bath is realized in terms of $N-1$ Ising models with the same spin configuration. It is convenient to choose such configuration as $\{\omega\}_1$, but, due to the arbitrariness of the chosen Ising model we can just denote by $\{\omega\}$ its spin configuration. Then we have

$$\prod_{j=1}^{N} \sum_{\{\omega\}_j \in \Omega} e^{-\eta \mathcal{E}(\{\omega\}_j)} = \mathcal{Z}^N(\eta),$$

where

$$\mathcal{Z}(\eta, J_1, J_2, H) := \sum_{\{\omega\} \in \Omega} e^{-\eta \mathcal{E}(\{\omega\})}.$$

• We can now write

$$\mathcal{Z}_{\mathrm{M}} = \frac{\Delta}{2\pi i} \int_{-i\pi/\Delta}^{+i\pi/\Delta} \exp\left(N\left(\eta \overline{E} + \log \mathcal{Z}(\eta, J_1, J_2, H) \right) \right) d\eta. \tag{4.10}$$

- The task is now to construct the asymptotic expansion for $N \gg 1$ of (4.10). We omit details of such expansion and just give the following (non-trivial!) claims:

  (a) The asymptotic value of (4.10) is attained at that value $\eta \in \mathbb{C}$ which solves the equation

  $$\overline{E} = -\frac{\partial}{\partial \eta} \log \mathcal{Z}(\eta, J_1, J_2, H). \qquad (4.11)$$

  (b) Equation (4.11) admits a unique solution for $\eta \in \mathbb{R}_+$.

  (c) The value $\eta \in \mathbb{R}_+$ which solves (4.11) can be identified with $\beta := (\kappa T)^{-1}$.

  (d) The asymptotic expansion of (4.10) can be written as

  $$
  \begin{aligned}
  \mathcal{Z}_M &= \frac{\Delta}{2\pi N^{1/2}} \exp\left(N\left(\beta\overline{E} + \log \mathcal{Z}(\beta, J_1, J_2, H)\right)\right) \\
  &\times \left(\int_{-\infty}^{+\infty} \exp\left(-\frac{x^2}{2}\frac{\partial^2}{\partial\beta^2}\log \mathcal{Z}(\beta, J_1, J_2, H)\right) dx\right)\left(1 + O\left(\frac{1}{N}\right)\right).
  \end{aligned}
  $$

  (e) Evaluating the above integral (see Lemma 2.1) one gets

  $$\mathcal{Z}_M \approx \Delta \frac{\mathcal{Z}^N(\beta, J_1, J_2, H)\, e^{N\beta\overline{E}}}{\left(2\pi N \dfrac{\partial^2}{\partial\beta^2}\log \mathcal{Z}(\beta, J_1, J_2, H)\right)^{1/2}}. \qquad (4.12)$$

- A similar asymptotic analysis gives the asymptotic expansion

  $$
  \begin{aligned}
  \sum_{\{w\}_2 \in \Omega} \cdots \sum_{\{w\}_N \in \Omega} \delta_{\mathcal{E}_{tot}, E_{tot}} &\approx \Delta \frac{\mathcal{Z}^{N-1}(\beta, J_1, J_2, H)\, e^{\beta N\overline{E}}\, e^{-\beta\mathcal{E}(\{w\}_1)}}{\left(2\pi N \dfrac{\partial^2}{\partial\beta^2}\log \mathcal{Z}(\beta, J_1, J_2, H)\right)^{1/2}} \\
  &= \mathcal{Z}_M \frac{e^{-\beta\mathcal{E}(\{w\}_1)}}{\mathcal{Z}(\beta, J_1, J_2, H)} \qquad (4.13)
  \end{aligned}
  $$

- By using the definition (4.5) we can just construct the ratio between (4.13) and (4.12). This gives the asymptotic formula (4.6).

The Theorem is proved. ∎

▶ Remarks:

- Note that, as one may expect, the probability (4.6) does not depend on the difference Δ between energy levels.

- Formula (4.7) derived for $\{w\}_1$, is valid for any any spin configuration $\{w\}_\ell$, $\ell = 1, \ldots, N$. Therefore we are allowed to drop the index 1.

- We define the *canonical partition function* of the Ising model with configurational energy (4.1) as

$$\mathcal{Z}(\beta, J_1, J_2, H) := \sum_{\{\omega\} \in \Omega} e^{-\beta \mathcal{E}(\{\omega\})}. \tag{4.14}$$

- The discrete probability distribution of the CE is given by

$$P(\{\omega\}, \beta, J_1, J_2, H) := \frac{e^{-\beta \mathcal{E}(\{\omega\}))}}{\mathcal{Z}(\beta, J_1, J_2, H)}.$$

- The CE-average of a random variable $f : \Omega \to \mathbb{R}$ is defined by

$$\langle f(\{\omega\}) \rangle := \frac{1}{\mathcal{Z}(\beta, J_1, J_2, H)} \sum_{\{\omega\} \in \Omega} f(\{\omega\}) e^{-\beta \mathcal{E}(\{\omega\})}.$$

- The above definition agrees with formula (4.8). The average total configurational energy $\overline{E} := E_{\text{tot}}/N$ is obtained as

$$\overline{E} := \langle \mathcal{E}(\{\omega\}) \rangle \quad := \quad \frac{1}{\mathcal{Z}(\beta, J_1, J_2, H)} \sum_{\{\omega\} \in \Omega} \mathcal{E}(\{\omega\}) e^{-\beta \mathcal{E}(\{\omega\})}$$

$$= \quad -\frac{\partial}{\partial \beta} \log \mathcal{Z}(\beta, J_1, J_2, H).$$

- The limit $\beta \gg 1$ in (4.14) gives

$$\mathcal{Z}(\beta, J_1, J_2, H) \approx g \, e^{-\beta \mathcal{E}_{\min}},$$

where $\mathcal{E}_{\min}$ is the minimum energy the system may attain and $g \in \mathbb{N}$ is the multiplicity of $\mathcal{E}_{\min}$. It is customary to associate temperature $T = 0$ with this minimum of energy.

Example 4.1 (*A three spins one-dimensional Ising model*)

Consider a one-dimensional Ising model consisting of three spins $\omega_i = \pm 1$, $i = 1, 2, 3$ (i.e. $(m, n) = (1, 3)$ or $(m, n) = (3, 1)$) and configurational energy

$$\mathcal{E}(\{\omega\}) := -J(\omega_1 \omega_2 + \omega_2 \omega_3) - H(\omega_1 + \omega_2 + \omega_3), \qquad J > 0, \qquad H > 0.$$

The list of all possible states of the system and the corresponding energies is given by

$$\begin{array}{ll} \uparrow\uparrow\uparrow & \mathcal{E}_1 = -2J - 3H, \\ \uparrow\downarrow\uparrow & \mathcal{E}_2 = 2J - H, \\ \uparrow\uparrow\downarrow & \mathcal{E}_3 = -H, \\ \downarrow\uparrow\uparrow & \mathcal{E}_4 = \mathcal{E}_3, \\ \downarrow\uparrow\downarrow & \mathcal{E}_5 = 2J + H, \\ \downarrow\downarrow\uparrow & \mathcal{E}_6 = H, \\ \uparrow\downarrow\downarrow & \mathcal{E}_7 = \mathcal{E}_6, \\ \downarrow\downarrow\downarrow & \mathcal{E}_8 = -2J + 3H. \end{array}$$

The canonical partition function reads

$$\mathcal{Z}(\beta, J, H) = \sum_{i=1}^{8} e^{-\beta \mathcal{E}_i}.$$

4.3.2 Thermodynamics and thermodynamic limit

▶ We now consider an Ising model, as defined in Section 4.2, with configurational energy (4.1). We can define its thermodynamics starting from the knowledge of the canonical partition function (4.14). The main thermodynamic quantities are here defined.

- The *average energy* is defined by

$$\overline{E} := -\frac{\partial}{\partial \beta} \log \mathcal{Z}(\beta, J_1, J_2, H). \tag{4.15}$$

- The *heat capacity* is defined by:

$$\overline{C} := \frac{\partial \overline{E}}{\partial T}. \tag{4.16}$$

- The *magnetization* is defined by

$$\overline{M} := \langle \omega_{\text{tot}} \rangle := \frac{1}{\mathcal{Z}(\beta, J_1, J_2, H)} \sum_{\{\omega\} \in \Omega} \omega_{\text{tot}} \, e^{-\beta \mathcal{E}(\{\omega\})}. \tag{4.17}$$

- The *magnetic susceptibility* is defined by

$$\overline{\chi} := \frac{\partial \overline{M}}{\partial H}. \tag{4.18}$$

- The *free energy* is defined by

$$\overline{F} := -\frac{1}{\beta} \log \mathcal{Z}(\beta, J_1, J_2, H).$$

Note that there is no reason, a priori, for $\overline{C}$ and $\overline{\chi}$ to exist for all T and H.

Example 4.2 (*A one spin Ising model*)

Consider the trivial case of a Ising model with one single spin, i.e. $(m, n) = (1, 1)$. In such a case the spin interacts only with the magnetic field.

- The partition function is:

$$\mathcal{Z}(\beta, H) = e^{\beta H} + e^{-\beta H} = 2\cosh(\beta H).$$

- From the partition function we obtain:

$$\overline{E} = -H\tanh(\beta H), \qquad \overline{M} = \tanh(\beta H), \qquad \overline{F} = -\frac{1}{\beta}\log\left(2\cosh(\beta H)\right).$$

Note that $\overline{E}$ is quite different from $\pm H$ and $\overline{M}$ is quite different from ± 1. This is not surprising since the Gibbs formalism provides meaningful physical results if the system is "large".

▶ In what follows we will be mainly interested in thermodynamic quantities under the TL $m \to +\infty$ and $n \to +\infty$, namely $\Lambda \to \mathbb{Z}^2$.

- When both m and n are large the total magnetization, internal energy and free energy will be, in general, proportional to $D := m\,n$.

- For each thermodynamic quantity defined above we can define the corresponding quantity under the TL. This will give us the definition of that quantity *per spin*, denoted by the same letter without overline. For example, the *free energy per spin* is defined by

$$F := \lim_{\substack{m \to +\infty \\ n \to +\infty}} \frac{\overline{F}}{m\,n}.$$

It is easy to see that the knowledge of F allows us to know other important thermodynamics quantities. In particular, the magnetization per spin can be written as

$$M = -\frac{\partial F}{\partial H}.$$

- The meaning of the TL can be understood looking at the canonical partition function (4.14):

    1. Let m and n be finite, so that Λ is a finite region of $\mathbb{Z}^2$.

        (a) The partition function (4.14) is a sum of a finite number of analytic functions of β and J_1, J_2, H and therefore it is analytic.

        (b) For a given $\beta > 0$ and $H \in \mathbb{R}$, the partition function (4.14) is a finite sum of positive numbers and hence it is positive.

        (c) The partition function (4.14) must be non-zero for some region where β is sufficiently close to the positive real axis and H is sufficiently close to the real axis. Therefore $\log \mathcal{Z}(\beta, J_1, J_2, H)$ (and hence the free energy per spin) must be an analytic function of β and H in this region.

    2. Let m and n tend to infinity.

(a) The partition function (4.14) is a sum of an infinite number of analytic functions of β and J_1, J_2, H, which, in principle, may diverge.

(b) The position of the zeros of the partition function (4.14) may converge to the positive β-axis or to the real H-axis and so in this limit $\log \mathcal{Z}(\beta, J_1, J_2, H)$ (and hence the free energy per spin) does not have to be an analytic function of $\beta > 0$ and $H \in \mathbb{R}$.

(c) These analyticity properties of the free energy correspond to qualitatively features that appear in the TL and which are not possible if m and n are finite. This is related with the mechanism of *phase transitions*.

▶ The special case $m = n = 1$ of Example 4.2 illustrates a general feature of "small" systems: one measurement of a quantity may differ substantially from its average computed in the CE (cf. Theorems 3.9 and 3.12).

Theorem 4.2

Consider a Ising model in the CE with configurational energy (4.1).

1. *Assume* $0 < C < +\infty$. *Under the TL* $m \to +\infty$ *and* $n \to +\infty$, *the quantity*

$$\frac{\langle \mathcal{E}^2(\{\omega\}) \rangle - \langle \mathcal{E}(\{\omega\}) \rangle^2}{m\,n}$$

goes to zero.

2. *Assume* $0 < \chi < +\infty$. *Under the TL* $m \to +\infty$ *and* $n \to +\infty$, *the quantity*

$$\frac{\langle \omega_{\text{tot}}^2 \rangle - \langle \omega_{\text{tot}} \rangle^2}{m\,n}$$

goes to zero.

Proof. We prove only the second claim. The first one can be proved in a similar way. We proceed by steps.

• Consider the identity

$$\mathcal{E}(\{\omega\}) = \mathcal{E}(\{\omega\})\big|_{H=0} - H\,\omega_{\text{tot}}.$$

• We want to compute $\langle \omega_{\text{tot}} \rangle$. Using the above identity we can write (4.17) as

$$\overline{M} := \langle \omega_{\text{tot}} \rangle := \frac{1}{\mathcal{Z}(\beta, J_1, J_2, H)} \sum_{\{\omega\} \in \Omega} \omega_{\text{tot}}\, e^{-\beta \mathcal{E}(\{\omega\})\big|_{H=0}}\, e^{\beta H \omega_{\text{tot}}}. \tag{4.19}$$

- We also have the following identity:

$$\frac{\partial \mathcal{Z}}{\partial H} = \beta \sum_{\{\omega\} \in \Omega} \left(-\frac{\partial \mathcal{E}}{\partial H} \right) e^{-\beta \mathcal{E}(\{\omega\})} = \beta \sum_{\{\omega\} \in \Omega} \omega_{\text{tot}} e^{-\beta \mathcal{E}(\{\omega\})},$$

so that

$$\overline{M} := \langle \omega_{\text{tot}} \rangle = \frac{1}{\beta} \frac{1}{\mathcal{Z}(\beta, J_1, J_2, H)} \frac{\partial \mathcal{Z}}{\partial H} = \frac{1}{\beta} \frac{\partial}{\partial H} \log \mathcal{Z}(\beta, J_1, J_2, H). \qquad (4.20)$$

- From (4.18), (4.19) and (4.20) we get

$$\begin{aligned}
\overline{\chi} &:= \frac{\partial \overline{M}}{\partial H} \\
&= \frac{\beta}{\mathcal{Z}(\beta, J_1, J_2, H)} \sum_{\{\omega\} \in \Omega} \omega_{\text{tot}}^2 e^{-\beta \mathcal{E}(\{\omega\})} \\
&\quad - \frac{1}{\mathcal{Z}^2(\beta, J_1, J_2, H)} \frac{\partial \mathcal{Z}}{\partial H} \sum_{\{\omega\} \in \Omega} \omega_{\text{tot}} e^{-\beta \mathcal{E}(\{\omega\})} \\
&= \beta \left\langle \omega_{\text{tot}}^2 \right\rangle - \beta \left\langle \omega_{\text{tot}} \right\rangle^2 .
\end{aligned}$$

- Therefore we have:

$$\frac{\langle \omega_{\text{tot}}^2 \rangle - \langle \omega_{\text{tot}} \rangle^2}{m \, n} = \frac{1}{\beta} \frac{\overline{\chi}}{m \, n},$$

namely

$$\frac{\langle \omega_{\text{tot}}^2 \rangle}{(m \, n)^2} - \frac{\overline{M}^2}{(m \, n)^2} = \frac{1}{\beta} \frac{\overline{\chi}}{(m \, n)^2}.$$

- In the TL, i.e., $m \, n \gg 1$, we get:

$$\frac{\langle \omega_{\text{tot}}^2 \rangle}{(m \, n)^2} - M^2 \approx \frac{\chi}{\beta} \frac{1}{m \, n},$$

where $0 < \chi < +\infty$. The last equation implies that a measurement of the magnetization per spin will almost certainly yield M.

The second claim is proved.   ∎

▶ We conclude our discussion on the thermodynamics of Ising models with the following claim.

Theorem 4.3

Consider a Ising model in the CE with configurational energy (4.1) and let m and

> *n be finite. Then the magnetization $\overline{M} = \overline{M}(H)$ is an odd function of H which is analytic for all $H \in \mathbb{R}$ and for all $\beta > 0$. In particular $\overline{M}(0) = 0$.*

Proof. We proceed by steps.

- We immediately notice that if $H = 0$, then neither spin $+1$ nor -1 is preferred. Thus $\overline{M}(0) = 0$.

- If $H \neq 0$ then symmetry (4.2) implies that

$$\overline{M}(H) = -\overline{M}(-H).$$

- Let $\mathscr{E}_0 := \mathscr{E}(\omega)\big|_{H=0}$. Denote by $\sum'_\omega$ the summation over all states such that

$$\sum_{(i,j)\in\Lambda} \omega_{i,j} > 0.$$

- Then we have:

$$
\begin{aligned}
\overline{M}(H) &= \frac{1}{\mathcal{Z}(\beta, J_1, J_2, H)} \sum_{\{\omega\}\in\Omega} \omega_{\text{tot}}\, e^{-\beta\mathscr{E}_0}\, e^{\beta H \omega_{\text{tot}}} \\
&= \frac{1}{\mathcal{Z}(\beta, J_1, J_2, H)} \sum_{\{\omega\}\in\Omega}{}' \omega_{\text{tot}}\, e^{-\beta\mathscr{E}_0} \left(e^{\beta H \omega_{\text{tot}}} - e^{-\beta H \omega_{\text{tot}}} \right),
\end{aligned}
$$

  from which there follows that $\overline{M}(H) > 0$ if $H > 0$.

- The fact that $\overline{M}$ is analytic for all $H \in \mathbb{R}$ and for all $\beta > 0$ follows from the analyticity of $\log \mathcal{Z}(\beta, J_1, J_2, H)$ and from formula (4.20). In particular, $\overline{M}$ is a continuous function of H at $H = 0$ for all $\beta > 0$, so that $\lim_{H\to 0+} \overline{M} = 0$.

The Theorem is proved. ∎

▶ In the TL we still have that $M(H) = -M(-H)$, but the magnetization per spin M does not have to be analytic (and continuous) at $H = 0$ for all β. In particular,

$$M(0^+) := \lim_{H\to 0+} M(H) \geqslant \lim_{\substack{m\to+\infty\\ n\to+\infty}} \lim_{H\to 0+} \frac{\overline{M}(H)}{m\,n} \geqslant 0.$$

The quantity $M(0^+)$ is called *spontaneous magnetization* and the temperature at which $M(0^+)$ starts to become positive is called *critical temperature*.

4.4 One-dimensional Ising model

▶ There is a substantial difference between one-dimensional and two-dimensional Ising models.

- In the one-dimensional case the nearest neighbors interaction assumption is not necessary to simplify calculations and to solve the problem. Furthermore, the model does not exhibit phase transitions.

- In the two-dimensional case the nearest neighbors interaction assumption is necessary to simplify calculations and to solve the problem. The model exhibits phase transitions.

▶ The one-dimensional Ising model is defined as a reduction of the two-dimensional Ising model defined in Section 4.2.

- Let $\Lambda \subset \mathbb{Z}$ be a discrete line with n sites.

- Λ can admit the following boundary conditions:

    1. *Free boundary conditions.* Λ is free, i.e., open.
    2. *Cyclic boundary conditions.* Λ is periodic, i.e., wrapped on a circle.

- At each site there is a spin $\omega_i = \pm 1, i = 1, \dots, n$. There is a total of 2^n possible configurations of spins on Λ, a configuration $\{\omega\}$ being specified by the n spin variables, i.e.,

$$\{\omega\} := \{\omega_i, i \in \Lambda\}.$$

Therefore the phase space of the system is

$$\Omega := \{+1, -1\}^n.$$

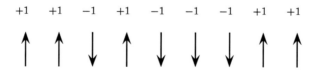

Fig. 4.3. A free one-dimensional Ising model with nine spins ([LaBe]).

- The system is in thermal equilibrium at temperature $T > 0$.

- We assume that only nearest neighbor spins can interact among each other. For a given configuration $\{\omega\} \in \Omega$ the configurational energy is the function

$$\mathcal{E}(\{\omega\}) := -\sum_{i \in \Lambda} (J\,\omega_i\,\omega_{i+1} + H\,\omega_i), \qquad (4.21)$$

where $J, H \in \mathbb{R}$. Such configurational energy can be written as

$$\mathscr{E}_f(\{\omega\}) := -J \sum_{i=1}^{n-1} \omega_i\, \omega_{i+1} - H \sum_{i=1}^{n} \omega_i, \tag{4.22}$$

in the case of a free lattice and

$$\mathscr{E}_c(\{\omega\}) := -J \sum_{i=1}^{n} \omega_i\, \omega_{i+1} - H \sum_{i=1}^{n} \omega_i, \qquad \omega_{n+1} \equiv \omega_1. \tag{4.23}$$

in the case of a periodic lattice.

▶ Our task is to compute the partition function for both energies (4.22) and (4.23). From the partition function we will derive the corresponding thermodynamics, thus showing that the free energy per spin does not depend on the boundary conditions and is an analytic function of the temperature.

4.4.1 Partition function

▶ To construct the canonical partition functions corresponding to configurational energies (4.22) and (4.23) we use the so called *transfer matrix method*, an algebraic technique which admits a generalization to the two-dimensional case.

▶ The transfer matrix method is based on the following construction.

- Consider spins ω_i and ω_{i+1}. If $\omega_i = \omega_{i+1}$ then they give a contribution $-J$ to the configurational energy, whereas if $\omega_i = -\omega_{i+1}$ the contribution is $+J$.

- With this pair of spins we associate the energy

$$\frac{1}{2} H \left(\omega_i + \omega_{i+1} \right).$$

- Let the two values which any element of $\{\omega\}$ may take to be the basis of a two-dimensional vector space. For any two values $\omega_i, \omega_j = \pm 1$ we define, in this vector space, a symmetric 2×2 matrix τ by setting

$$\tau_{\omega_i, \omega_j} \equiv \left\langle \omega_i \left| \tau \right| \omega_j \right\rangle := \exp\left(\beta J \omega_i \omega_j + \beta \frac{H}{2} \left(\omega_i + \omega_j \right) \right).$$

In other words,

$$\tau_{+1,+1} \equiv \left\langle +1 \left| \tau \right| +1 \right\rangle := e^{\beta\,(J+H)},$$
$$\tau_{+1,-1} \equiv \left\langle +1 \left| \tau \right| -1 \right\rangle := e^{-\beta J},$$
$$\tau_{-1,+1} \equiv \left\langle -1 \left| \tau \right| +1 \right\rangle := e^{-\beta J},$$
$$\tau_{-1,-1} \equiv \left\langle -1 \left| \tau \right| -1 \right\rangle := e^{\beta\,(J-H)},$$

namely

$$\tau := \begin{pmatrix} e^{\beta\,(J+H)} & e^{-\beta J} \\ e^{-\beta J} & e^{\beta\,(J-H)} \end{pmatrix}.$$ (4.24)

The matrix (4.24) is called *transfer matrix*.

(a) Note that

$$\sum_{\omega_i=\pm 1} \langle\, \omega_i \mid \tau \mid \omega_i \,\rangle = \text{Trace } \tau.$$

(b) Note the following identity

$$\sum_{\omega_k=\pm 1} \langle\, \omega_j \mid \tau \mid \omega_k \,\rangle \langle\, \omega_k \mid \tau \mid \omega_\ell \,\rangle = \exp\left(\beta\,(\omega_j+\omega_\ell)\left(\frac{H}{2}+J\right)+\beta H\right)$$

$$+ \exp\left(\beta\,(\omega_j+\omega_\ell)\left(\frac{H}{2}-J\right)-\beta H\right).$$

The same result is obtained by computing $\langle\, \omega_j \mid \tau^2 \mid \omega_\ell \,\rangle$. Indeed,

$$\tau^2 = \begin{pmatrix} e^{2\beta\,(J+H)}+e^{-2\beta J} & e^{\beta H}+e^{-\beta H} \\ e^{\beta H}+e^{-\beta H} & e^{2\beta\,(J-H)}+e^{-2\beta J} \end{pmatrix},$$

which confirms that

$$\sum_{\omega_k=\pm 1} \langle\, \omega_j \mid \tau \mid \omega_k \,\rangle \langle\, \omega_k \mid \tau \mid \omega_\ell \,\rangle = \langle\, \omega_j \mid \tau^2 \mid \omega_\ell \,\rangle.$$

This result can be easily generalized. For instance, there holds

$$\sum_{\omega_k=\pm 1}\sum_{\omega_\ell=\pm 1} \langle\, \omega_j \mid \tau \mid \omega_k \,\rangle \langle\, \omega_k \mid \tau \mid \omega_\ell \,\rangle \langle\, \omega_\ell \mid \tau \mid \omega_s \,\rangle = \langle\, \omega_j \mid \tau^3 \mid \omega_s \,\rangle.$$

(c) Since τ is symmetric it can be diagonalized by a similarity transform with some 2×2 matrix $\mathbf{A}$,

$$\tau \mapsto \mathbf{A}^{-1}\,\tau\,\mathbf{A} = \text{diag}(\lambda_+,\lambda_-),$$ (4.25)

where $\lambda_\pm$ satisfy the algebraic equation

$$\left(e^{\beta\,(J+H)}-\lambda\right)\left(e^{\beta\,(J-H)}-\lambda\right)-e^{-2\beta J}=0.$$

Its solutions are:

$$\lambda_\pm := e^{\beta J}\left(\cosh(\beta H) \pm \sqrt{\sinh^2(\beta H)+e^{-4\beta J}}\right).$$ (4.26)

Note that when $H \in \mathbb{R}$ and $\beta > 0$ we have $\lambda_+ > \lambda_-$.

(d) A possible choice for **A** is

$$\mathbf{A} := \begin{pmatrix} -e^{\beta J}\left(e^{\beta(J-H)} - \lambda_+\right) & 1 \\ 1 & -e^{\beta J}\left(e^{\beta(J+H)} - \lambda_-\right) \end{pmatrix}. \tag{4.27}$$

Note that

$$\mathbf{A}^{-1} = -\mathbf{A}\,\mathrm{diag}\left(\frac{1}{\det \mathbf{A}}, \frac{1}{\det \mathbf{A}}\right).$$

(e) Note that

$$\mathrm{Trace}\,\tau = \mathrm{Trace}\left(\mathbf{A}^{-1}\,\tau\,\mathbf{A}\right) = \lambda_+ + \lambda_-,$$

where we used the cyclic property of the trace. Moreover

$$\mathrm{Trace}\,\tau^n = \lambda_+^n + \lambda_-^n,$$

for all $n \in \mathbb{N}$.

- We define the two-dimensional vector v by setting

$$v_{\omega_i} \equiv \langle \omega_i \,|\, v \rangle = \left\langle v^\top \,\middle|\, \omega_i \right\rangle := e^{\beta H \omega_i / 2}.$$

Explicitly,

$$v = (v_{+1}, v_{-1}) := \left(e^{\beta H/2}, e^{-\beta H/2}\right)^\top.$$

It is easy to see that the following identity holds:

$$\sum_{\omega_j = \pm 1}\sum_{\omega_k = \pm 1} \left\langle v^\top \,\middle|\, \omega_j \right\rangle \langle \omega_j \,|\, \tau \,|\, \omega_k \rangle \langle \omega_k \,|\, v \rangle = v^\top \,\tau\, v.$$

▶ The next Theorem gives the explicit form of the canonical partition function of the one-dimensional Ising model.

Theorem 4.4

Consider a one-dimensional Ising model with configurational energy (4.21).

1. The canonical partition function in the case of free boundary conditions is

$$\mathcal{Z}_f(\beta, J, H) = \lambda_+^{n-1}\mu_+ + \lambda_-^{n-1}\mu_-, \tag{4.28}$$

where $\lambda_\pm$ are defined in (4.26) and

$$\mu_\pm := \cosh(\beta H) \pm \frac{\sinh^2(\beta H) + e^{-2\beta J}}{\sqrt{\sinh^2(\beta H) + e^{-4\beta J}}}. \tag{4.29}$$

2. *The canonical partition function in the case of cyclic boundary conditions is*

$$Z_c(\beta, J, H) = \lambda_+^n \left(1 + \left(\frac{\lambda_-}{\lambda_+} \right)^n \right), \tag{4.30}$$

where $\lambda_\pm$ are defined in (4.26).

Proof. We prove both results.

1. We have

$$Z_f(\beta, J, H) := \sum_{\omega_1 = \pm 1} \cdots \sum_{\omega_n = \pm 1} \exp \left(\beta J \sum_{i=1}^{n-1} \omega_i \omega_{i+1} + \beta H \sum_{i=1}^{n} \omega_i \right). \tag{4.31}$$

Note that the first spin ω_1 interacts only with the second spin ω_2 and with the magnetic field. Similarly, the last spin ω_n interacts only with ω_{n-1} and with the magnetic field. Therefore, by using the transfer matrix τ and the vector v we can write (4.31) as

$$
\begin{aligned}
Z_f &= \sum_{\omega_1 = \pm 1} \cdots \sum_{\omega_n = \pm 1} \left\langle v^\top \middle| \omega_1 \right\rangle \left\langle \omega_1 \middle| \tau \middle| \omega_2 \right\rangle \cdots \left\langle \omega_{n-1} \middle| \tau \middle| \omega_n \right\rangle \left\langle \omega_n \middle| v \right\rangle \\
&= \sum_{\omega_1 = \pm 1} \sum_{\omega_n = \pm 1} \left\langle v^\top \middle| \omega_1 \right\rangle \left\langle \omega_1 \middle| \tau^{n-1} \middle| \omega_n \right\rangle \left\langle \omega_n \middle| v \right\rangle \\
&= v^\top \tau^{n-1} v.
\end{aligned}
$$

Using (4.25) and the explicit form of **A** given in (4.27) we obtain the desired formula.

2. We have

$$Z_c(\beta, J, H) := \sum_{\omega_1 = \pm 1} \cdots \sum_{\omega_n = \pm 1} \exp \left(\beta J \sum_{i=1}^{n} \omega_i \omega_{i+1} + \beta H \sum_{i=1}^{n} \omega_i \right), \tag{4.32}$$

where $\omega_{n+1} \equiv \omega_1$. By using the transfer matrix τ we can write (4.32) as

$$
\begin{aligned}
Z_c &= \sum_{\omega_1 = \pm 1} \cdots \sum_{\omega_n = \pm 1} \left\langle \omega_1 \middle| \tau \middle| \omega_2 \right\rangle \cdots \left\langle \omega_n \middle| \tau \middle| \omega_1 \right\rangle \\
&= \sum_{\omega_1 = \pm 1} \left\langle \omega_1 \middle| \tau^n \middle| \omega_1 \right\rangle \\
&= \text{Trace } \tau^n,
\end{aligned}
$$

which gives the desired formula.

The Theorem is proved.   ∎

▶ Assume that $H = 0$. Then from Theorem 4.4 we observe that:

- The partition functions (4.28) and (4.30) reduce respectively to

$$\mathcal{Z}_f(\beta, J) = 2 \left(2\cosh(\beta J)\right)^{n-1},$$
$$\mathcal{Z}_c(\beta, J) = \left(2\cosh(\beta J)\right)^{n} \left(1 + \tanh^{n}(\beta J)\right),$$

- Note that $\mathcal{Z}_f$ is an even function of J whereas $\mathcal{Z}_c$ is an even function of J if n is even but possesses no such symmetry if n is odd. These features may be understood by looking at the configurational energies $\mathcal{E}_f$ and $\mathcal{E}_c$ and by considering the replacement $\omega_k \mapsto (-1)^k \omega'_k$. For instance, one has:

$$\mathcal{E}_f(\{\omega\}, J) \mapsto \mathcal{E}_f(\{\omega'\}, -J),$$

which guarantees that $\mathcal{Z}_f$ is an even function of J.

- Consider the limit $\beta \gg 1$. We have:

$$\mathcal{Z}_f(\beta, J) \approx 2\, e^{\beta |J|(n-1)},$$

so that $(\mathcal{E}_f)_{\min} = -|J|(n-1)$ with degeneracy 2. On the other hand we have:

$$\mathcal{Z}_c(\beta, J) \approx \begin{cases} 2\, e^{\beta |J| n} & \text{if } J > 0 \text{ or if } J < 0, \ n \text{ even,} \\ 2n\, e^{\beta |J|(n-2)} & \text{if } J < 0, \ n \text{ odd.} \end{cases}$$

- The difference between the two cases can be explained on the basis on the following figure, where the configuration of minimum of energy is considered when $J < 0$ and $H = 0$ in the case of cyclic boundary conditions (*antiferromagnetic regime*).

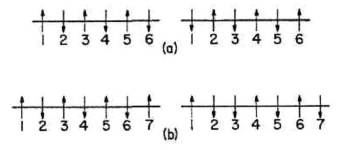

Fig. 4.4. Free/cyclic one-dimensional Ising models with a even/odd number of spins ([McCWu]).

When n is even both configurations are shown in (a). The two ground states have the same energy. When n is odd (figure (b)) we show only two of the $2n$ configurations. The regular alternation of spins, $\omega_k \mapsto (-1)^k \omega'_k$, must be broken at one bond. At this bond spins must be both either "up" (+1) or "down" (−1). This phenomenon appears in the one-dimensional Ising model only at $T = 0$, while it occurs also at some critical temperature for the two-dimensional Ising model.

4.4.2   Thermodynamics

▶ The thermodynamics of the one-dimensional Ising model is constructed under the TL, i.e., $n \rightarrow +\infty$. The next claim shows that, as one may expect, the free energy per spin does not depend on the boundary conditions.

Theorem 4.5

In the TL $n \rightarrow +\infty$ the free energy per site F of a one-dimensional Ising model does not depend on the boundary conditions. We have:

$$F = -\frac{1}{\beta} \lim_{n \to +\infty} \frac{1}{n} \log \mathcal{Z}_f(\beta, J, H) = -\frac{1}{\beta} \lim_{n \to +\infty} \frac{1}{n} \log \mathcal{Z}_c(\beta, J, H) = -\frac{1}{\beta} \log \lambda_+.$$

Explicitly,

$$F = -J - \frac{1}{\beta} \log \left(\cosh(\beta H) + \sqrt{\sinh^2(\beta H) + e^{-4\beta J}} \right).$$

Proof. We have $\lambda_+ > \lambda_-$ (for $\beta > 0$). Thus, from the partition function $\mathcal{Z}_c$, we find, for $n \gg 1$,

$$\log \mathcal{Z}_c(\beta, J, H) = n \log \lambda_+ + O\left(\frac{\lambda_-^n}{\lambda_+^n} \right).$$

From the partition function $\mathcal{Z}_f$, we find, for $n \gg 1$,

$$\log \mathcal{Z}_f(\beta, J, H) = n \log \lambda_+ + \log \lambda_+ - \log \mu_+ + O\left(\frac{\lambda_-^n}{\lambda_+^n} \right).$$

Therefore both $n^{-1} \log \mathcal{Z}_c$ and $n^{-1} \log \mathcal{Z}_f$ approach, for $n \gg 1$, $\log \lambda_+$. The claim follows. ■

▶ Starting from the knowledge of the partition function we can derive not only the free energy, but also other thermodynamic quantities.

• The heat capacity per spin is

$$C := -\kappa \beta^2 \frac{\partial^2 F}{\partial \beta^2},$$

which turns out to be a quite complicated expression of β, J, H. However it is a differentiable function of H for $\beta \geqslant 0$. A qualitative plot of C against T is given below.

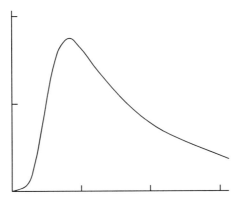

Fig. 4.5. Heat capacity per spin plotted against the temperature ([LaBe]).

If $H = 0$ the quantity C simplifies to

$$C = \kappa \beta^2 J^2 \text{sech}^2(\beta J).$$

- The magnetization per spin is

$$M = -\frac{\partial F}{\partial H} = \frac{\sinh(\beta H)}{\sqrt{\sinh^2(\beta H) + e^{-4\beta J}}},$$

which is a monotone and differentiable function of H for $\beta \geqslant 0$. What this result suggests is that there is no spontaneous magnetization: if $H = 0$ the magnetization vanishes, even in the TL.

▶ We conclude with the following observations:

- For finite n both $\mathcal{Z}_c$ and $\mathcal{Z}_f$ are analytic functions of β and H for all $\beta \geqslant 0$ and H.

- If $n \to +\infty$ there are values of β and H where neither the partition function nor the free energy spin is analytic. This lack on analyticity, however, does not occur for $H \subset \mathbb{R}$ and $\beta \geqslant 0$, so that there is no occurence of phase transitions.

4.5 Two-dimensional Ising model

▶ We consider a two-dimensional Ising model, as defined in Section 4.2, with toroidal boundary conditions, $m = n$ (i.e., $D = n^2$) and $J_1 = J_2 =: J$. Such choice does not influence the derivation of the thermodynamics under the TL.

▶ We will see that the construction of the partition function, using the transfer matrix method, will be much more involved than the same construction for the one-dimensional model. This is a dramatic effect of dimensionality: in the one-dimensional case we had only to find the largest eigenvalue of a 2×2 matrix. In the two-dimensional case we have to find the largest eigenvalue of a $2^n \times 2^n$ matrix. Our presentation is based on the classical paper by B. Kaufman, "Crystal Statistics II. Partition Function Evaluated by Spinor Analysis", Phys. Rev. 76/8, 1949.

▶ The first task is to find a formal expression for the canonical partition function corresponding to the configurational energy (see (4.1))

$$\mathscr{E}(\{\omega\}) := -J \sum_{i,j=1}^{n} \left(\omega_{i,j}\,\omega_{i,j+1} + \omega_{i,j}\,\omega_{i+1,j} \right) - H \sum_{i,j=1}^{n} \omega_{i,j}, \qquad (4.33)$$

where $\omega_{i,n+1} \equiv \omega_{i,1}$ and $\omega_{n+1,j} \equiv \omega_{1,j}$.

- Recall that each spin variable $\omega_{i,j}$ can be parametrized as $\omega_\alpha, \alpha = 1, \ldots, n^2$. Let $\mu_\nu, \nu = 1, \ldots, n$, be the collection of all spin variables $\omega_\alpha, \alpha = 1, \ldots, n$, on the ν-th row:

$$\mu_\nu := \{\omega_1, \ldots, \omega_n\}_{\nu-\text{th row}}, \qquad \mu_{n+1} \equiv \mu_1.$$

  Note that there are a total of 2^n possible configurations for each row.

- Any configuration $\{\omega\} \in \Omega$ is specified by assigning n rows μ_ν, i.e., $\{\omega\} = \{\mu_1, \ldots, \mu_n\}$. If we do not need to specify the index of the row we simply write a row spin configuration as $\mu := \{\omega_1, \ldots, \omega_n\}$. Its neighboring row will be denoted by $\mu' := \{\omega'_1, \ldots, \omega'_n\}$.

- Each row interacts only with the neighboring rows and with the magnetic field. Therefore we define the local interaction energy between two neighboring rows as

$$\mathscr{E}_1(\mu, \mu') := -J \sum_{\alpha=1}^{n} \omega_\alpha\,\omega'_\alpha, \qquad (4.34)$$

  and the local interaction energy between spins within a given row plus their interaction with the magnetic field as

$$\mathscr{E}_2(\mu) := -J \sum_{\alpha=1}^{n} \omega_\alpha\,\omega_{\alpha+1} - H \sum_{\alpha=1}^{n} \omega_\alpha, \qquad \omega_{n+1} \equiv \omega_1. \qquad (4.35)$$

- We can write the configurational energy (4.33) by summing over all rows $\nu = 1, \ldots, n$ the local energies (4.34) and (4.35):

$$\mathscr{E}(\{\omega\}) = \sum_{\nu=1}^{n} \left(\mathscr{E}_1(\mu_\nu, \mu_{\nu+1}) + \mathscr{E}_2(\mu_\nu) \right), \qquad \mu_{n+1} \equiv \mu_1 \qquad (4.36)$$

Note that (4.36) is formally the configurational energy of a periodic one-dimensional Ising model where single spins are replaced by rows of spins.

▶ The configurational energy (4.36) is now written in a way which allows us to introduce a transfer matrix.

- For notational convenience we introduce the parameters

$$\varepsilon := \beta J, \qquad h := \beta H.$$

- Define a $2^n \times 2^n$ matrix τ by setting

$$
\begin{aligned}
\langle \mu \,|\, \tau \,|\, \mu' \rangle &:= \exp\left(-\beta \mathscr{E}_1(\mu, \mu') - \beta \mathscr{E}_2(\mu)\right) \\
&= \prod_{\alpha=1}^{n} e^{\varepsilon \omega_\alpha \omega_{\alpha+1}} \, e^{\varepsilon \omega_\alpha \omega'_\alpha} \, e^{h \omega_\alpha}.
\end{aligned}
\tag{4.37}
$$

Such matrix is diagonalizable. Since the trace of a matrix is independent of the representation of the matrix, we can write

$$\text{Trace } \tau = \sum_{\alpha=1}^{2^n} \lambda_\alpha,$$

where λ_α, $\alpha = 1, \ldots, 2^n$, are the eigenvalues of τ.

- A computation similar to the one performed in the one-dimensional case shows that we can easily write the canonical partition function corresponding to (4.36) as the trace of the n-th power of the transfer matrix (see proof of Theorem 4.4):

$$
\begin{aligned}
\mathcal{Z}(\beta, J, H) &:= \sum_{\mu_1} \cdots \sum_{\mu_n} \exp\left(-\beta \sum_{\nu=1}^{n} \left(\mathscr{E}_1(\mu_\nu, \mu_{\nu+1}) + \mathscr{E}_2(\mu_\nu)\right)\right) \\
&= \sum_{\mu_1} \cdots \sum_{\mu_n} \langle \mu_1 \,|\, \tau \,|\, \mu_2 \rangle \cdots \langle \mu_n \,|\, \tau \,|\, \mu_1 \rangle \\
&= \sum_{\mu_1} \langle \mu_1 \,\big|\, \tau^n \,\big|\, \mu_1 \rangle \\
&= \text{Trace } \tau^n \\
&= \sum_{\alpha=1}^{2^n} (\lambda_\alpha)^n.
\end{aligned}
$$

▶ If $\lambda_{\max}$ is the largest eigenvalue of τ we expect that the limit

$$\lim_{n \to +\infty} \frac{1}{n} \log \lambda_{\max} \tag{4.38}$$

is a finite number.

- If this is true and if all eigenvalues λ_α are positive then

$$(\lambda_{\max})^n \leqslant \mathcal{Z}(\beta, J, H) \leqslant 2^n (\lambda_{\max})^n,$$

which implies

$$\frac{1}{n} \log \lambda_{\max} \leqslant \frac{1}{n^2} \log \mathcal{Z}(\beta, J, H) \leqslant \frac{1}{n} \log \lambda_{\max} + \frac{1}{n} \log 2.$$

- Therefore

$$\lim_{D \to +\infty} \frac{1}{D} \log \mathcal{Z}(\beta, J, H) = \lim_{n \to +\infty} \frac{1}{n} \log \lambda_{\max}.$$

- It will turn out that the limit (4.38) is finite and that all eigenvalues of τ are positive.

▶ In the remaining part of the Section we will spend some efforts to derive an explicit representation of τ. This will allow us to understand the structure of the spectrum of τ and therefore to compute the free energy per spin under the TL.

4.5.1 *Some algebraic tools: spinor analysis*

▶ The main algebraic problem is the diagonalization of τ. It will be useful to introduce some notions, definitions and claims (without proof).

- *Direct product of matrices.* Let **A** and **B** be two $m \times m$ matrices whose matrix elements are

$$\mathbf{A}_{ij} \equiv \langle i | \mathbf{A} | j \rangle, \qquad \mathbf{B}_{ij} \equiv \langle i | \mathbf{B} | j \rangle, \qquad i, j = 1, \dots, m.$$

  (a) The *direct product* (or *Kronecker product*) between **A** and **B** is the $m^2 \times m^2$ matrix, denoted by $\mathbf{A} \otimes \mathbf{B}$, whose entries are

$$\langle ik | \mathbf{A} \otimes \mathbf{B} | j\ell \rangle := \langle i | \mathbf{A} | j \rangle \langle k | \mathbf{B} | \ell \rangle, \qquad i, j, k, \ell = 1, \dots, m.$$

  This definition can be extended to the direct product of any number of $m \times m$ matrices.

  (b) If **A**, **B**, **C**, **D** are $m \times m$ matrices there holds

$$(\mathbf{A} \otimes \mathbf{B})(\mathbf{C} \otimes \mathbf{D}) = \mathbf{A}\,\mathbf{C} \otimes \mathbf{B}\,\mathbf{D}.$$

  This formula can be extended to any arbitrary number of $m \times m$ matrices.

- *Pauli matrices.* The *Pauli matrices* are 2×2 matrices defined by

$$\sigma^1 := \begin{pmatrix} 0 & 1 \\ 1 & 0 \end{pmatrix}, \qquad \sigma^2 := \begin{pmatrix} 0 & -i \\ i & 0 \end{pmatrix}, \qquad \sigma^3 := \begin{pmatrix} 1 & 0 \\ 0 & -1 \end{pmatrix}.$$

(a) The following formulas are true:

$$(\sigma^i)^2 = \mathbf{1}, \qquad \sigma^i \sigma^j + \sigma^j \sigma^i = \mathbf{0}, \qquad \sigma^i \sigma^j = i \sigma^k,$$

and

$$e^{\theta \sigma^i} = \cosh \theta + \sigma^i \sinh \theta, \tag{4.39}$$

where (i, j, k) is any cyclic permutation of $(1, 2, 3)$ and $\theta \in \mathbb{C}$.

(b) By using the direct product we define the following $2^n \times 2^n$ matrices:

$$\sigma^i_\alpha := \underbrace{\mathbf{1} \otimes \cdots \otimes \mathbf{1} \otimes \sigma^i \otimes \mathbf{1} \otimes \cdots \otimes \mathbf{1}}_{n \text{ factors}}, \qquad i = 1, 2, 3,$$

where σ^i is the α-th factor. One easily proves that

$$\left[\sigma^i_\alpha, \sigma^i_\beta \right] = \mathbf{0}, \qquad \left[\sigma^i_\alpha, \sigma^j_\beta \right] = \mathbf{0}, \qquad \alpha \neq \beta,$$

where the bracket denotes the standard commutator. Moreover, there holds

$$e^{\theta \sigma^i_\alpha} = \cosh \theta + \sigma^i_\alpha \sinh \theta,$$

where $\theta \in \mathbb{C}$.

- *Generalized Dirac matrices.* Let $\{\gamma_\alpha\}$ be a set of $2n$ matrices defined by the anti-commutation rule

$$\gamma_\alpha \gamma_\beta + \gamma_\beta \gamma_\alpha = 2 \delta_{\alpha\beta} \mathbf{1}, \qquad \alpha, \beta = 1, \ldots, 2n. \tag{4.40}$$

Such matrices are called *generalized Dirac matrices* (or *gamma matrices*). The following statements are true:

(a) $d := \dim \gamma_\alpha \geqslant 2^n \times 2^n$.

(b) Recall that the *general linear Lie group* on $\mathbb{C}$ is defined by

$$\mathbf{GL}(N, \mathbb{C}) := \{ \mathbf{S} \in \mathfrak{gl}(N, \mathbb{C}) \ : \ \det \mathbf{S} \neq 0 \}.$$

Here $\mathfrak{gl}(N, \mathbb{C})$ is the *general linear Lie algebra*, i.e., the vector space of all linear maps (not necessarily invertible) from $\mathbb{C}^N$ to $\mathbb{C}^N$. Therefore, $\mathbf{GL}(N, \mathbb{C})$ is the non-commutative group of all invertible $N \times N$ matrices (with coefficients in $\mathbb{C}$), where the group operation is given by the usual product of matrices. Now, if $\{\gamma_\alpha\}, \{\gamma'_\alpha\}, \alpha = 1, \ldots, 2n$, satisfy (4.40) then there exists $\mathbf{S} \in \mathbf{GL}(\sqrt{d}, \mathbb{C})$ such that

$$\gamma_\alpha = \mathbf{S} \gamma'_\alpha \mathbf{S}^{-1}$$

for all $\alpha = 1, \ldots, 2n$.

(c) Any d-dimensional complex square matrix is a linear combination of the unit matrix, the matrices $\{\gamma_\alpha\}$ and all independent products between the matrices $\{\gamma_\alpha\}$. Also the converse is true.

(d) A $(2^n \times 2^n)$-dimensional representation of the set $\{\gamma_\alpha\}$ is given by

$$
\begin{aligned}
\gamma_{2\alpha-1} &:= \sigma_1^1 \sigma_2^1 \cdots \sigma_{\alpha-1}^1 \sigma_\alpha^3, \\
\gamma_{2\alpha} &:= \sigma_1^1 \sigma_2^1 \cdots \sigma_{\alpha-1}^1 \sigma_\alpha^2,
\end{aligned}
$$

where $\alpha = 1, \ldots, n$. In this representation one has the formulas

$$
\gamma_{2\alpha}\,\gamma_{2\alpha-1} = i\,\sigma_\alpha^1, \qquad \gamma_{2\alpha+1}\,\gamma_{2\alpha} = i\,\sigma_\alpha^3\,\sigma_{\alpha+1}^3, \tag{4.41}
$$

and

$$
\gamma_1\,\gamma_{2n} = -i\,\sigma_1^3\,\sigma_n^3 \prod_{\alpha=1}^{n} \sigma_\alpha^1. \tag{4.42}
$$

Hereafter we work with the above representation of $\{\gamma_\alpha\}$. Therefore d is fixed to $2^n \times 2^n$.

(e) Define the matrix

$$
\mathbf{U} := \prod_{\alpha=1}^{n} \sigma_\alpha^1. \tag{4.43}
$$

The following formulas hold true:

$$
\mathbf{U}^2 = \mathbf{1}, \qquad \mathbf{U}\,(\mathbf{1} \pm \mathbf{U}) = \pm\mathbf{1} + \mathbf{U}, \qquad \mathbf{U} = i^n \prod_{\alpha=1}^{2n} \gamma_\alpha, \tag{4.44}
$$

and

$$
\begin{aligned}
\exp\left(i\theta\,\gamma_1\,\gamma_{2n}\,\mathbf{U}\right) &= \frac{1+\mathbf{U}}{2}\,\exp\left(i\theta\,\gamma_1\,\gamma_{2n}\right) \\
&\quad + \frac{1-\mathbf{U}}{2}\,\exp\left(-i\theta\,\gamma_1\,\gamma_{2n}\right),
\end{aligned} \tag{4.45}
$$

where $\theta \in \mathbb{R}$. Note that $\mathbf{U}$ can be expressed as the product of an even number of generalized Dirac matrices. Therefore, as a consequence of (4.40), $\mathbf{U}$ commutes with any product of an even number of matrices γ_α and anticommutes with any product of an odd number of matrices γ_α.

- *Orthogonal group and its spin representation.* Let

$$
\mathbf{O}(N,\mathbb{C}) := \left\{ \mathbf{W} \in \mathbf{GL}(N,\mathbb{C}) : \mathbf{W}^\top \mathbf{W} = \mathbf{1} \right\}
$$

be the matrix Lie group of complex orthogonal matrices, which is a subgroup of $\mathbf{GL}(N,\mathbb{C})$. The dimension of $\mathbf{O}(N,\mathbb{C})$ is $N(N-1)$ (over $\mathbb{R}$) and the action of $\mathbf{O}(N,\mathbb{C})$ on $\mathbb{C}^N$ defines a *rotation*.

(a) Consider a set of matrices $\{\gamma_\alpha\}$, $\alpha = 1, \ldots, 2n$, and define another set $\{\gamma'_\alpha\}$ by setting

$$\gamma' := \mathbf{W}\,\gamma, \qquad \mathbf{W} \in \mathbf{O}(2n, \mathbb{C}),$$

where $\gamma := (\gamma_1, \ldots, \gamma_{2n})^\top$ and $\gamma' := (\gamma'_1, \ldots, \gamma'_{2n})^\top$. This explicitly means that

$$\gamma'_\alpha := \sum_{\beta=1}^{2n} \mathbf{W}_{\alpha\beta}\,\gamma_\beta, \qquad \alpha = 1, \ldots, 2n. \tag{4.46}$$

It turns out that also the matrices $\{\gamma'_\alpha\}$ satisfy (4.40) and, as a consequence, there exists $\mathbf{S_W} \in \mathbf{GL}(2^n, \mathbb{C})$ such that

$$\gamma'_\alpha = \mathbf{S_W}\,\gamma_\alpha\,\mathbf{S_W^{-1}}, \qquad \alpha = 1, \ldots, 2n. \tag{4.47}$$

From (4.46) and (4.47) there follows that

$$\mathbf{S_W}\,\gamma_\alpha\,\mathbf{S_W^{-1}} = \sum_{\beta=1}^{2n} \mathbf{W}_{\alpha\beta}\,\gamma_\beta, \qquad \alpha = 1, \ldots, 2n. \tag{4.48}$$

One says that $\mathbf{S_W}$ is a *spin representative* of $\mathbf{W}$. One can prove that if $\mathbf{S}_{\mathbf{W}_1}$ and $\mathbf{S}_{\mathbf{W}_2}$ satisfy (4.48) then

$$\mathbf{S}_{\mathbf{W}_1\mathbf{W}_2} = \mathbf{S}_{\mathbf{W}_1}\,\mathbf{S}_{\mathbf{W}_2}, \qquad \mathbf{W}_1, \mathbf{W}_2 \in \mathbf{O}(2n, \mathbb{C}).$$

(b) We consider some special rotations, namely (planar) rotations in the plane $\alpha\beta$ ($\alpha \neq \beta$) by an angle $\theta \in \mathbb{R}$. They are defined by matrices $\mathbf{W}(\alpha, \beta \,|\, \theta) \in \mathbf{O}(2n, \mathbb{C})$ whose non-vanishing entries are

$$\mathbf{W}_{\nu\nu}(\alpha, \beta \,|\, \theta) \quad := \quad 1, \tag{4.49}$$
$$\mathbf{W}_{\alpha\alpha}(\alpha, \beta \,|\, \theta) = \mathbf{W}_{\beta\beta}(\alpha, \beta \,|\, \theta) \quad := \quad \cos\theta, \tag{4.50}$$
$$\mathbf{W}_{\alpha\beta}(\alpha, \beta \,|\, \theta) = \mathbf{W}_{\beta\alpha}(\alpha, \beta \,|\, \theta) \quad := \quad -\sin\theta, \tag{4.51}$$

where $\nu, \alpha, \beta = 1, \ldots, 2n$ and $\nu \neq \alpha, \beta$. Note that

$$(\mathbf{W}(\alpha, \beta \,|\, \theta))^{-1} = \mathbf{W}(\alpha, \beta \,|\, -\theta) = \mathbf{W}(\beta, \alpha \,|\, \theta) = (\mathbf{W}(\alpha, \beta \,|\, \theta))^\top. \tag{4.52}$$

According to (4.46), we have:

$$\gamma'_\nu \quad := \quad \gamma_\nu,$$
$$\gamma'_\alpha \quad := \quad \gamma_\alpha \cos\theta - \gamma_\beta \sin\theta,$$
$$\gamma'_\beta \quad := \quad \gamma_\alpha \sin\theta + \gamma_\beta \cos\theta,$$

where $\nu, \alpha, \beta = 1, \ldots, 2n$ and $\nu \neq \alpha, \beta$. One can prove that:

1. The eigenvalues of $\mathbf{W}(\alpha, \beta \,|\, \theta)$ are 1, each $(2n - 2)$-fold degenerate, and $e^{\pm i\theta}$ (non-degenerate).

2. Consider the product of n commuting planar rotations of type (4.49–4.51), with angles $\theta_1, \ldots, \theta_n \in \mathbb{R}$:

$$\prod_{i=1}^{n} \mathbf{W}(\alpha_i, \beta_i \mid \theta_i), \tag{4.53}$$

where $(\alpha_1, \beta_1, \ldots, \alpha_n, \beta_n)$ is a permutation of $(1, 2, \ldots, 2n-1, 2n)$. Then the $2n$ eigenvalues of (4.53) are

$$e^{\pm i\theta_1}, \ldots, e^{\pm i\theta_n}.$$

3. The spin representative of $\mathbf{W}(\alpha, \beta \mid \theta)$, say $\mathbf{S}_{\mathbf{W}(\alpha, \beta \mid \theta)}$, satisfies (4.48) and admits the representation

$$\mathbf{S}_{\mathbf{W}(\alpha, \beta \mid \theta)} = \exp\left(-\frac{\theta}{2}\, \gamma_\alpha\, \gamma_\beta\right),$$

whose eigenvalues are

$$e^{\pm i\theta/2},$$

each 2^{n-1}-fold degenerate.

4. The spin representative of (4.53), say $\mathbf{S}_{\prod_{i=1}^{n} \mathbf{W}(\alpha_i, \beta_i \mid \theta_i)}$, satisfies (4.48) and admits the representation

$$\mathbf{S}_{\prod_{i=1}^{n} \mathbf{W}(\alpha_i, \beta_i \mid \theta_i)} = \exp\left(-\frac{\theta_1}{2}\, \gamma_{\alpha_1}\, \gamma_{\beta_1}\right) \cdots \exp\left(-\frac{\theta_n}{2}\, \gamma_{\alpha_n}\, \gamma_{\beta_n}\right),$$

which has 2^n eigenvalues given by

$$e^{i(\pm\theta_1 \pm \cdots \pm \theta_n)/2}.$$

Here the signs $\pm$ are to be chosen independently.

4.5.2   *Algebraic structure of the transfer matrix*

▶ The above algebraic tools will allow us to understand the structure of the transfer matrix τ of the two-dimensional Ising model. Recall that our boundary conditions are toroidal. We start with the following result.

Lemma 4.1

The transfer matrix (4.37) can be written as

$$\tau = \mathbf{V}_3\, \mathbf{V}_2\, \mathbf{V}_1',$$

where $\mathbf{V}_1', \mathbf{V}_2, \mathbf{V}_3$ *are* $2^n \times 2^n$ *matrices defined by*

$$\langle \mu \,|\, \mathbf{V}_1' \,|\, \mu' \rangle \quad := \quad \prod_{\alpha=1}^{n} e^{\varepsilon \, \omega_\alpha \, \omega_\alpha'}, \tag{4.54}$$

$$\langle \mu \,|\, \mathbf{V}_2 \,|\, \mu' \rangle \quad := \quad \delta_{\omega_1 \omega_1'} \cdots \delta_{\omega_n \omega_n'} \prod_{\alpha=1}^{n} e^{\varepsilon \, \omega_\alpha \, \omega_{\alpha+1}}, \tag{4.55}$$

$$\langle \mu \,|\, \mathbf{V}_3 \,|\, \mu' \rangle \quad := \quad \delta_{\omega_1 \omega_1'} \cdots \delta_{\omega_n \omega_n'} \prod_{\alpha=1}^{n} e^{h \, \omega_\alpha}. \tag{4.56}$$

In particular, if $H = 0$ *then* $\mathbf{V}_3 = \mathbf{1}$ *and*

$$\tau \big|_{H=0} = \mathbf{V}_2 \, \mathbf{V}_1'$$

Proof. Recall that $\mu := \{\omega_1, \ldots, \omega_n\}$ denotes a row spin configuration and $\mu' := \{\omega_1', \ldots, \omega_n'\}$ is its neighboring row. It is then obvious that in the usual sense of matrix multiplication we can write

$$\begin{aligned}
\langle \mu \,|\, \tau \,|\, \mu' \rangle \quad &= \quad \prod_{\alpha=1}^{n} e^{\varepsilon \, \omega_\alpha \, \omega_{\alpha+1}} \, e^{\varepsilon \, \omega_\alpha \, \omega_\alpha'} \, e^{h \, \omega_\alpha} \\
&= \quad \sum_{\mu''} \sum_{\mu'''} \langle \mu \,|\, \mathbf{V}_3 \,|\, \mu'' \rangle \langle \mu'' \,|\, \mathbf{V}_2 \,|\, \mu''' \rangle \langle \mu''' \,|\, \mathbf{V}_1' \,|\, \mu' \rangle,
\end{aligned}$$

where the matrix entries appearing in the sums are defined as in (4.54–4.56). This proves the claim. ∎

▶ The next lemma allows us to represent the matrices $\mathbf{V}_1', \mathbf{V}_2, \mathbf{V}_3$ in terms of direct products of Pauli matrices.

Lemma 4.2

The matrices $\mathbf{V}_1', \mathbf{V}_2, \mathbf{V}_3$ *can be written as*

$$\begin{aligned}
\mathbf{V}_1' \quad &= \quad (2 \sinh(2\varepsilon))^{n/2} \, \mathbf{V}_1, \\
\mathbf{V}_2 \quad &= \quad \prod_{\alpha=1}^{n} \exp\left(\varepsilon \, \sigma_\alpha^3 \, \sigma_{\alpha+1}^3 \right), \qquad \sigma_{n+1}^3 \equiv \sigma_1^3, \\
\mathbf{V}_3 \quad &= \quad \prod_{\alpha=1}^{n} \exp\left(h \, \sigma_\alpha^3 \right),
\end{aligned}$$

where

$$\mathbf{V}_1 := \prod_{\alpha=1}^{n} \exp\left(\theta \, \sigma_\alpha^1 \right), \qquad \tanh \theta := e^{-2\varepsilon}.$$

Proof. We prove explicitly the result only for $\mathbf{V}_1'$. From the expression (4.54) it is clear that $\mathbf{V}_1'$ is the direct product of n 2×2 identical matrices of the form

$$\mathbf{A} := \begin{pmatrix} e^\varepsilon & e^{-\varepsilon} \\ e^{-\varepsilon} & e^\varepsilon \end{pmatrix} = e^\varepsilon \mathbf{1} + e^{-\varepsilon} \sigma^1.$$

Using (4.39) we get

$$\mathbf{A} = (2\sinh(2\varepsilon))^{1/2} \exp\left(\theta\,\sigma^1\right),$$

with θ defined by $\tanh\theta := e^{-2\varepsilon}$. Therefore we have

$$\begin{aligned} \mathbf{V}_1' &= (2\sinh(2\varepsilon))^{n/2} \underbrace{\exp\left(\theta\,\sigma^1\right) \otimes \cdots \otimes \exp\left(\theta\,\sigma^1\right)}_{n \text{ factors}} \\ &= (2\sinh(2\varepsilon))^{n/2} \prod_{\alpha=1}^{n} \exp\left(\theta\,\sigma_\alpha^1\right), \end{aligned}$$

which is the desired formula. A similar computation gives the expressions for $\mathbf{V}_2$ and $\mathbf{V}_3$. ∎

▶ From Lemmas 4.1 and 4.2 the next claim follows.

Theorem 4.6

> *The transfer matrix (4.37) can be written as*
>
> $$\tau = (2\sinh(2\varepsilon))^{n/2}\,\mathbf{V}_3\,\mathbf{V}_2\,\mathbf{V}_1.$$

▶ We will be interested in computing the partition function of the two-dimensional Ising model without external magnetic field, i.e., $H = 0$. Therefore, we are interested in the structure of the transfer matrix τ when $\mathbf{V}_3 = \mathbf{1}$. We know from Theorem 4.6 that in such a case the transfer matrix has the form

$$\tau = (2\sinh(2\varepsilon))^{n/2}\,\mathbf{V}_2\,\mathbf{V}_1.$$

▶ Hereafter we assume $H = 0$. The next lemma allows us to represent the matrices $\mathbf{V}_1, \mathbf{V}_2$ in terms of generalized Dirac matrices.

Lemma 4.3

> *The matrices $\mathbf{V}_1, \mathbf{V}_2$ can be written as*
>
> $$\begin{aligned} \mathbf{V}_1 &= \prod_{\alpha=1}^{n} \exp\left(-\mathrm{i}\,\theta\,\gamma_{2\alpha}\,\gamma_{2\alpha-1}\right), \\ \mathbf{V}_2 &= \exp\left(\mathrm{i}\,\varepsilon\,\mathbf{U}\,\gamma_1\,\gamma_{2n}\right) \prod_{\alpha=1}^{n-1} \exp\left(-\mathrm{i}\,\varepsilon\,\gamma_{2\alpha+1}\,\gamma_{2\alpha}\right), \end{aligned}$$

where **U** *is the matrix defined in (4.43).*

Proof. The expression for $\mathbf{V}_1$ follows from the first formula (4.41). The expression for $\mathbf{V}_2$ can be easily obtained:

$$
\begin{aligned}
\mathbf{V}_2 &= \prod_{\alpha=1}^{n} \exp\left(\varepsilon\, \sigma_\alpha^3\, \sigma_{\alpha+1}^3\right) = \exp\left(\varepsilon\, \sigma_n^3\, \sigma_1^3\right) \prod_{\alpha=1}^{n-1} \exp\left(\varepsilon\, \sigma_\alpha^3\, \sigma_{\alpha+1}^3\right) \\
&= \exp\left(i\varepsilon\, \mathbf{U}\, \gamma_1\, \gamma_{2n}\right) \prod_{\alpha=1}^{n-1} \exp\left(-i\varepsilon\, \gamma_{2\alpha+1}\, \gamma_{2\alpha}\right),
\end{aligned}
$$

where we used the second formula (4.41), formula (4.42) and (4.43). ∎

▶ We now define the matrix $\mathbf{V} := \mathbf{V}_2\, \mathbf{V}_1$, so that

$$
\tau = \left(2\sinh(2\,\varepsilon)\right)^{n/2} \mathbf{V}. \tag{4.57}
$$

- It will be shown that $\mathbf{V}$ can be expressed in terms of 2^n-dimensional spin representatives of $2n$-dimensional rotations and that, as a consequence, its eigenvalues are known as soon as the eigenvalues of the rotations are determined.

- Let us recall that we are interested in the largest eigenvalue of τ, $\lambda_{\max}$. More precisely, taking also into account (4.57), we are interested in the computation of the limit

$$
\begin{aligned}
\lim_{D \to +\infty} \frac{1}{D} \log \mathcal{Z}(\beta, \varepsilon, 0) &= \lim_{n \to +\infty} \frac{1}{n} \log \lambda_{\max} \\
&= \frac{1}{2} \log\left(2\sinh(2\,\varepsilon)\right) + \lim_{n \to +\infty} \frac{1}{n} \log \widetilde{\lambda}_{\max}, \tag{4.58}
\end{aligned}
$$

where $\widetilde{\lambda}_{\max}$ denotes the largest eigenvalue of $\mathbf{V}$. We also recall that (4.58) is meaningful if and only if all eigenvalues of $\mathbf{V}$ are positive and the limit exists and is finite.

▶ Lemma 4.3 allows us to prove the next claim.

Theorem 4.7

The matrix $\mathbf{V}$ *can be written as*

$$
\mathbf{V} = \frac{1+\mathbf{U}}{2}\, \mathbf{S}_+ + \frac{1-\mathbf{U}}{2}\, \mathbf{S}_-, \tag{4.59}
$$

where

$$
\mathbf{S}_\pm := \exp\left(\pm i\varepsilon\, \gamma_1\, \gamma_{2n}\right) \prod_{\alpha=1}^{n-1} \exp\left(-i\varepsilon\, \gamma_{2\alpha+1}\, \gamma_{2\alpha}\right) \prod_{\alpha=1}^{n} \exp\left(-i\theta\, \gamma_{2\alpha}\, \gamma_{2\alpha-1}\right). \tag{4.60}
$$

Proof. From Lemma 4.3 and formula (4.45) we get

$$
\mathbf{V}_2 = \frac{1+\mathbf{U}}{2} \exp\left(\mathrm{i}\,\varepsilon\,\gamma_1\,\gamma_{2n}\right) \prod_{\alpha=1}^{n-1} \exp\left(-\mathrm{i}\,\varepsilon\,\gamma_{2\alpha+1}\,\gamma_{2\alpha}\right)
$$
$$
+\frac{1-\mathbf{U}}{2} \exp\left(-\mathrm{i}\,\varepsilon\,\gamma_1\,\gamma_{2n}\right) \prod_{\alpha=1}^{n-1} \exp\left(-\mathrm{i}\,\varepsilon\,\gamma_{2\alpha+1}\,\gamma_{2\alpha}\right),
$$

so that

$$
\mathbf{V} = \frac{1+\mathbf{U}}{2} \exp\left(\mathrm{i}\,\varepsilon\,\gamma_1\,\gamma_{2n}\right) \prod_{\alpha=1}^{n-1} \exp\left(-\mathrm{i}\,\varepsilon\,\gamma_{2\alpha+1}\,\gamma_{2\alpha}\right) \prod_{\alpha=1}^{n} \exp\left(-\mathrm{i}\,\theta\,\gamma_{2\alpha}\,\gamma_{2\alpha-1}\right)
$$
$$
+\frac{1-\mathbf{U}}{2} \exp\left(-\mathrm{i}\,\varepsilon\,\gamma_1\,\gamma_{2n}\right) \prod_{\alpha=1}^{n-1} \exp\left(-\mathrm{i}\,\varepsilon\,\gamma_{2\alpha+1}\,\gamma_{2\alpha}\right) \prod_{\alpha=1}^{n} \exp\left(-\mathrm{i}\,\theta\,\gamma_{2\alpha}\,\gamma_{2\alpha-1}\right),
$$

which is the claim.   ∎

▶ It is now evident that the matrices $\mathbf{S}_\pm$ given in (4.60) are spin representatives of suitable rotations.

4.5.3   The case $H = 0$. Diagonalization of the transfer matrix

▶ Our task is to diagonalize the transfer matrix $\tau = \left(2\sinh(2\,\varepsilon)\right)^{n/2}\mathbf{V}$, where $\mathbf{V}$ is expressed in terms of formula (4.59). We will see that we can restrict our attention to the diagonalization of the matrices $\mathbf{S}_\pm$.

▶ We start with the following observations.

- A set of diagonalizable matrices commutes if and only if the set is simultaneously diagonalizable. The three diagonalizable matrices $\mathbf{U}$ and $\mathbf{S}_\pm$ are the building blocks of the the transfer matrix τ (see (4.59)). For a fixed n the matrix $\mathbf{U}$ is composed by $2n$ generalized Dirac matrices (see (4.44)), while $\mathbf{S}_\pm$ are products of $2n$ exponentials of generalized Dirac matrices where each exponential contains a product of two generalized Dirac matrices (see (4.60)). Therefore, as a consequence of (4.40), the matrices $\mathbf{U}$ and $\mathbf{S}_\pm$ are a set of commuting matrices and they can be simultaneously diagonalized.

- We first consider a similarity transformation on $\mathbf{V}$, induced by a matrix $\mathbf{R} \in GL(2^n, \mathbb{C})$, which diagonalizes $\mathbf{U}$ but not necessarily $\mathbf{S}_\pm$:

$$
\tilde{\mathbf{V}} := \mathbf{R}\,\mathbf{V}\,\mathbf{R}^{-1} = \frac{1+\tilde{\mathbf{U}}}{2}\,\tilde{\mathbf{S}}_+ + \frac{1-\tilde{\mathbf{U}}}{2}\,\tilde{\mathbf{S}}_-,
$$

where

$$
\tilde{\mathbf{U}} := \mathbf{R}\,\mathbf{U}\,\mathbf{R}^{-1}, \qquad \tilde{\mathbf{S}}_\pm := \mathbf{R}\,\mathbf{S}_\pm\,\mathbf{R}^{-1}.
$$

- Since $\mathbf{U}^2 = \mathbf{1}$, the 2^n eigenvalues of $\mathbf{U}$ are ± 1 and eigenvalues $+1$ and -1 occur with the same frequency 2^{n-1}. We can choose $\mathbf{R}$ in such a way that the diagonal matrix $\widetilde{\mathbf{U}}$ has the form

$$\widetilde{\mathbf{U}} = \mathrm{diag}(\mathbf{1}, -\mathbf{1}),$$

where $\mathbf{1}$ is a $2^{n-1} \times 2^{n-1}$ identity matrix.

- Since $\widetilde{\mathbf{S}}_\pm$ commute with $\widetilde{\mathbf{U}}$ they must have the form

$$\widetilde{\mathbf{S}}_\pm = \mathrm{diag}(\mathbf{A}_\pm, \mathbf{B}_\pm),$$

where $\mathbf{A}_\pm$ and $\mathbf{B}_\pm$ are $2^{n-1} \times 2^{n-1}$ matrices not necessarily diagonal.

- We now note that

$$\frac{1 + \widetilde{\mathbf{U}}}{2} \widetilde{\mathbf{S}}_+ = \mathrm{diag}(\mathbf{A}_+, \mathbf{0}), \qquad \frac{1 - \widetilde{\mathbf{U}}}{2} \widetilde{\mathbf{S}}_- = \mathrm{diag}(\mathbf{0}, \mathbf{B}_-),$$

so that

$$\widetilde{\mathbf{V}} = \mathrm{diag}(\mathbf{A}_+, \mathbf{B}_-). \tag{4.61}$$

- To diagonalize $\mathbf{V}$ it is sufficient to diagonalize $\widetilde{\mathbf{V}}$, which has the same set of eigenvalues of $\mathbf{V}$. Hence, from (4.61), we see that we need to diagonalize the matrices $\mathrm{diag}(\mathbf{A}_+, \mathbf{0})$ and $\mathrm{diag}(\mathbf{0}, \mathbf{B}_-)$ separately and independently. The combined set of their non-vanishing eigenvalues is the set of eigenvalues of $\mathbf{V}$.

- To diagonalize $\mathrm{diag}(\mathbf{A}_+, \mathbf{0})$ and $\mathrm{diag}(\mathbf{0}, \mathbf{B}_-)$ we should diagonalize $\widetilde{\mathbf{S}}_\pm$ separately and independently, thus obtaining twice too many eigenvalues for each (i.e., eigenvalues of $\mathbf{A}_-$ and $\mathbf{B}_+$). To obtain the eigenvalues of $\mathrm{diag}(\mathbf{A}_+, \mathbf{0})$ and $\mathrm{diag}(\mathbf{0}, \mathbf{B}_-)$ we need to decide which eigenvalues are to be discarded. This last step is however not necessary if we consider the TL since we will be interested only in the largest eigenvalue of $\mathbf{V}$.

- Instead of diagonalizing $\widetilde{\mathbf{S}}_\pm$ we will diagonalize $\mathbf{S}_\pm$, which have the same set of eigenvalues of $\widetilde{\mathbf{S}}_\pm$.

▶ The next Theorem provides the eigenvalues of the matrices $\mathbf{S}_\pm$. As anticipated, the problem of diagonalizing $\mathbf{S}_\pm$ is solved by finding the eigenvalues of the $2n$-dimensional rotations for which $\mathbf{S}_\pm$ are spin representatives.

Theorem 4.8

1. *The rotation matrices* $\mathbf{W}_\pm \in \mathbf{O}(2n, \mathbb{C})$ *corresponding to* $\mathbf{S}_\pm$ *are given by*

$$\mathbf{W}_\pm = \mathbf{W}(1, 2n \mid \mp 2\mathrm{i}\varepsilon) \prod_{\alpha=1}^{n-1} \mathbf{W}(2\alpha + 1, 2\alpha \mid -2\mathrm{i}\varepsilon) \prod_{\alpha=1}^{n} \mathbf{W}(2\alpha, 2\alpha - 1 \mid -2\mathrm{i}\theta).$$

2. The matrix $\mathbf{W}_+$ has $2n$ positive eigenvalues $e^{\pm\ell_k}$, $k = 1, 3, \ldots, 2n-1$, where the ℓ_k's solve the equation

$$\cosh \ell_k = \cosh(2\,\varepsilon) \cosh(2\,\theta) - \cos\left(\frac{\pi k}{n}\right) \sinh(2\,\varepsilon) \sinh(2\,\theta), \qquad (4.62)$$

with $k = 1, 3, \ldots, 2n-1$.

3. The matrix $\mathbf{W}_-$ has $2n$ positive eigenvalues $e^{\pm\ell_k}$, $k = 0, 2, \ldots, 2n-2$, where the ℓ_k's solve the equation (4.62) with $k = 0, 2, \ldots, 2n-2$.

4. For $k = 0, 1, \ldots, 2n-1$, one has $\ell_k = \ell_{2n-k}$ and

$$0 < \ell_0 < \ell_1 < \cdots < \ell_n.$$

5. The matrix $\mathbf{S}_+$ has 2^n eigenvalues given by

$$e^{(\pm\ell_1 \pm \ell_3 \pm \ell_5 \pm \cdots \pm \ell_{2n-1})/2}. \qquad (4.63)$$

Here all possible choices of the signs $\pm$ are to be made independently.

6. The matrix $\mathbf{S}_-$ has 2^n eigenvalues given by

$$e^{(\pm\ell_0 \pm \ell_2 \pm \ell_4 \pm \cdots \pm \ell_{2n-2})/2}. \qquad (4.64)$$

Here all possible choices of the signs $\pm$ are to be made independently.

Proof. We prove all claims. We proceed by steps.

- From the expressions (4.60) it is clear that $\mathbf{S}_\pm$ are 2^n-dimensional spin representatives of products of planar rotations acting on a $2n$-dimensional space (see (4.53)). The rotation matrices corresponding to $\mathbf{S}_\pm$ are given by

$$\mathbf{W}_\pm = \mathbf{W}(1, 2n \,|\, \mp 2\,\mathrm{i}\,\varepsilon) \prod_{\alpha=1}^{n-1} \mathbf{W}(2\,\alpha+1, 2\,\alpha \,|\, -2\,\mathrm{i}\,\varepsilon) \prod_{\alpha=1}^{n} \mathbf{W}(2\,\alpha, 2\,\alpha-1 \,|\, -2\,\mathrm{i}\,\theta).$$

This proves the first claim.

- We are interested in computing the eigenvalues of $\mathbf{W}_\pm$, which are the same as those of the matrices

$$\widetilde{\mathbf{W}}_\pm := \mathbf{D}\,\mathbf{W}_\pm\,\mathbf{D}^{-1},$$

where

$$\mathbf{D} := \prod_{\alpha=1}^{n} \mathbf{W}(2\,\alpha, 2\,\alpha-1 \,|\, -\mathrm{i}\,\theta) = \prod_{\alpha=1}^{n} \mathbf{W}(2\,\alpha-1, 2\,\alpha \,|\, \mathrm{i}\,\theta).$$

Note that, using (4.52) we get

$$\mathbf{D}^{-1} = \prod_{\alpha=1}^{n} \mathbf{W}(2\alpha, 2\alpha - 1 \,|\, i\theta) = \prod_{\alpha=1}^{n} \mathbf{W}(2\alpha - 1, 2\alpha \,|\, -i\theta).$$

- Explicitly we have:

$$
\begin{aligned}
\widetilde{\mathbf{W}}_\pm &= \mathbf{W}(1,2, \,|\, i\theta) \cdots \mathbf{W}(2n-1, 2n \,|\, i\theta) \\
&\quad \mathbf{W}(1, 2n \,|\, \mp 2 i\varepsilon)\, \mathbf{W}(2,3 \,|\, 2 i\varepsilon) \cdots \mathbf{W}(2n-2, 2n-1 \,|\, 2 i\varepsilon) \\
&\quad \mathbf{W}(1,2, \,|\, 2 i\theta) \cdots \mathbf{W}(2n-1, 2n \,|\, 2 i\theta) \\
&\quad \mathbf{W}(1,2, \,|\, -i\theta) \cdots \mathbf{W}(2n-1, 2n \,|\, -i\theta) \\
&= \mathbf{W}(1,2, \,|\, i\theta) \cdots \mathbf{W}(2n-1, 2n \,|\, i\theta) \\
&\quad \mathbf{W}(1, 2n \,|\, \mp 2 i\varepsilon)\, \mathbf{W}(2,3 \,|\, 2 i\varepsilon) \cdots \mathbf{W}(2n-2, 2n-1 \,|\, 2 i\varepsilon) \\
&\quad \mathbf{W}(1,2, \,|\, i\theta) \cdots \mathbf{W}(2n-1, 2n \,|\, i\theta), \tag{4.65}
\end{aligned}
$$

where we used (4.52) and the fact that rotations acting on different planes commute, while rotations acting on the same plane combine.

- From (4.65) we see that

$$\widetilde{\mathbf{W}}_\pm := \mathbf{D}\,\mathbf{C}_\pm\,\mathbf{D}, \tag{4.66}$$

where

$$\mathbf{C}_\pm := \mathbf{W}(1, 2n \,|\, \mp 2 i\varepsilon) \prod_{\alpha=1}^{n-1} \mathbf{W}(2\alpha, 2\alpha + 1 \,|\, 2 i\varepsilon).$$

- Explicitly, $\mathbf{D}$ and $\mathbf{C}_\pm$ are $2n \times 2n$ matrices given by

$$
\mathbf{D} := \begin{pmatrix}
\cosh\theta & i\sinh\theta & 0 & 0 & \cdots & 0 & 0 \\
-i\sinh\theta & \cosh\theta & 0 & 0 & \cdots & 0 & 0 \\
0 & 0 & \cosh\theta & i\sinh\theta & \cdots & 0 & 0 \\
0 & 0 & -i\sinh\theta & \cosh\theta & \cdots & 0 & 0 \\
\vdots & \vdots & \vdots & \vdots & \vdots & \vdots & \vdots \\
0 & 0 & 0 & 0 & \cdots & \cosh\theta & i\sinh\theta \\
0 & 0 & 0 & 0 & \cdots & -i\sinh\theta & \cosh\theta
\end{pmatrix},
$$

and

$$
\mathbf{C}_\pm := \begin{pmatrix}
\cosh(2\varepsilon) & 0 & 0 & \cdots & 0 & 0 & \pm i\sinh(2\varepsilon) \\
0 & \cosh(2\varepsilon) & i\sinh(2\varepsilon) & \cdots & 0 & 0 & 0 \\
0 & -i\sinh(2\varepsilon) & \cosh(2\varepsilon) & \cdots & 0 & 0 & 0 \\
\vdots & \vdots & \vdots & \vdots & \vdots & \vdots & \vdots \\
0 & 0 & 0 & \cdots & \cosh(2\varepsilon) & i\sinh(2\varepsilon) & 0 \\
0 & 0 & 0 & \cdots & -i\sinh(2\varepsilon) & \cosh(2\varepsilon) & 0 \\
\mp i\sinh(2\varepsilon) & 0 & 0 & \cdots & 0 & 0 & \cosh(2\varepsilon)
\end{pmatrix}.
$$

- Performing the matrix multiplication (4.66) we obtain

$$
\widetilde{W}_\pm := \begin{pmatrix}
\mathbf{a} & \mathbf{b} & 0 & 0 & \cdots & 0 & \mp\mathbf{b}^* \\
\mathbf{b}^* & \mathbf{a} & \mathbf{b} & 0 & \cdots & 0 & 0 \\
0 & \mathbf{b}^* & \mathbf{a} & \mathbf{b} & \cdots & 0 & 0 \\
\vdots & \vdots & \vdots & \vdots & \vdots & \vdots & \vdots \\
0 & 0 & 0 & 0 & \cdots & \mathbf{a} & \mathbf{b} \\
\mp\mathbf{b} & 0 & 0 & 0 & \cdots & \mathbf{b}^* & \mathbf{a}
\end{pmatrix},
$$

  where

$$
\mathbf{a} := \begin{pmatrix} \cosh(2\varepsilon)\cosh(2\theta) & -i\cosh(2\varepsilon)\sinh(2\theta) \\ i\cosh(2\varepsilon)\sinh(2\theta) & \cosh(2\varepsilon)\cosh(2\theta) \end{pmatrix}, \tag{4.67}
$$

$$
\mathbf{b} := -\frac{1}{2} \begin{pmatrix} \sinh(2\varepsilon)\sinh(2\theta) & -2i\sinh(2\varepsilon)\sinh^2\theta \\ 2i\sinh(2\varepsilon)\cosh^2\theta & \sinh(2\varepsilon)\sinh(2\theta) \end{pmatrix}, \tag{4.68}
$$

  and $\mathbf{b}^*$ is the Hermitian conjugate of $\mathbf{b}$.

- To find the eigenvalues of $\widetilde{W}_\pm$ we make the following Ansatz for its eigenvectors:

$$
\psi := (z\,\phi, z^2\,\phi, \ldots, z^n\,\phi)^\top,
$$

  where $z \in \mathbb{C}$ and $\phi := (\phi_1, \phi_2)$ is a two-component vector. The eigenvector problem

$$
\widetilde{W}_\pm \psi = \lambda\,\psi,
$$

  where λ is one of the eigenvalues of $\widetilde{W}_\pm$, leads to the following eigenvalues equations:

$$
\left(z\,\mathbf{a} + z^2\,\mathbf{b} \mp z^n\,\mathbf{b}^* \right) \phi = z\,\lambda\,\phi,
$$

$$
\left(z^2\,\mathbf{a} + z^3\,\mathbf{b} + z\,\mathbf{b}^* \right) \phi = z^2\,\lambda\,\phi,
$$

$$
\left(z^3\,\mathbf{a} + z^4\,\mathbf{b} + z^2\,\mathbf{b}^* \right) \phi = z^3\,\lambda\,\phi,
$$

$$
\vdots
$$

$$
\left(z^{n-1}\,\mathbf{a} + z^n\,\mathbf{b} + z^{n-2}\,\mathbf{b}^* \right) \phi = z^{n-1}\,\lambda\,\phi,
$$

$$
\left(z^n\,\mathbf{a} \mp z\,\mathbf{b} + z^{n-1}\,\mathbf{b}^* \right) \phi = z^n\,\lambda\,\phi.
$$

- Only three of the above eigenvalues equations are independent: the first one, the last one and any one between the first and the last one, namely,

$$
\left(\mathbf{a} + z\,\mathbf{b} \mp z^{n-1}\,\mathbf{b}^* \right) \phi = \lambda\,\phi, \tag{4.69}
$$

$$
\left(\mathbf{a} + z\,\mathbf{b} + z^{-1}\,\mathbf{b}^* \right) \phi = \lambda\,\phi, \tag{4.70}
$$

$$
\left(\mathbf{a} \mp z^{1-n}\,\mathbf{b} + z^{-1}\,\mathbf{b}^* \right) \phi = \lambda\,\phi. \tag{4.71}
$$

- Note that (4.69–4.71) are solved by putting $z^n = \mp 1$. Then (4.69–4.71) reduce to a single eigenvalue equation, say (4.70).

- Therefore for $\widetilde{\mathbf{W}}_+$ and $\widetilde{\mathbf{W}}_-$ there are n values of z given by

$$z_k = e^{(2 i \pi k)/n}, \qquad k = 0, 1, \ldots, 2n - 1,$$

where

$$k = \begin{cases} 0, 2, 4, \ldots, 2n - 2 & \text{for } \widetilde{\mathbf{W}}_+, \\ 1, 3, 5, \ldots, 2n - 1 & \text{for } \widetilde{\mathbf{W}}_-. \end{cases} \tag{4.72}$$

For each k there are two eigenvalues λ_k of $\widetilde{\mathbf{W}}_\pm$:

$$\left(\mathbf{a} + z_k \, \mathbf{b} + z_k^{-1} \, \mathbf{b}^* \right) \phi = \lambda_k \, \phi,$$

where λ_k is associated with $\widetilde{\mathbf{W}}_\pm$ according to (4.72).

- We now need to find explicitly λ_k. From (4.67) and (4.68) we see that

$$\det \mathbf{a} = 1, \qquad \det \mathbf{b} = \det \mathbf{b}^* = 0, \qquad \det \left(\mathbf{a} + z_k \, \mathbf{b} + z_k^{-1} \, \mathbf{b}^* \right) = 1,$$

which imply that the two values of λ_k must have the form

$$\lambda_k = e^{\pm \ell_k}, \qquad k = 0, 1, \ldots, 2n - 1, \tag{4.73}$$

where the ℓ_k's solve the equation

$$\frac{1}{2} \operatorname{Trace} \left(\mathbf{a} + z_k \, \mathbf{b} + z_k^{-1} \, \mathbf{b}^* \right) = \frac{1}{2} \left(e^{\ell_k} + e^{-\ell_k} \right) = \cosh \ell_k. \tag{4.74}$$

The trace on the l.h.s. may be directly evaluated by using (4.67) and (4.68). One finds that (4.74) reduces to (4.62):

$$\cosh(2\varepsilon) \cosh(2\theta) - \cos \left(\frac{\pi k}{n} \right) \sinh(2\varepsilon) \sinh(2\theta) = \cosh \ell_k. \tag{4.75}$$

It is clear that if ℓ_k is a solution of (4.75) then also $-\ell_k$ is a solution. But this possibility has already been taken into account in (4.73). The second and the third claim are proved.

- We prove the fourth claim. The fact that $\ell_k = \ell_{2n-k}$ follows from a direct check on (4.75). The fact that $0 < \ell_0 < \ell_1 < \cdots < \ell_n$ can be proved by noticing that

$$\frac{\partial \ell_k}{\partial k} = \frac{\pi}{n} \sin \left(\frac{\pi k}{n} \right) \frac{1}{\sin \ell_k},$$

which is positive for $k \leqslant n$. A plot of ℓ_k as a function of ε is given in the figure below.

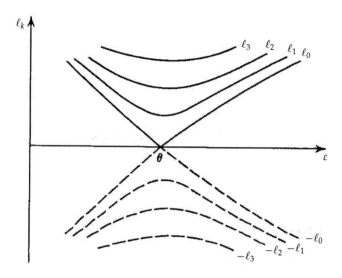

Fig. 4.6. Solutions of (4.75) [Hu]

- The last two claims immediately follow from our general considerations on spin representatives of rotation matrices.

The Theorem is proved.   ■

▶ As we anticipated, the set of eigenvalues of $\mathbf{V}$ consists of one-half the set of eigenvalues of $\mathbf{S}_+$ and one-half that of $\mathbf{S}_-$. Therefore, to find explicitly the eigenvalues of $\mathbf{V}$ we must decide which half of the set of eigenvalues of $\mathbf{S}_\pm$ can be discarded. However, this step is not necessary if we consider our problem under the TL, since we are interested only in the largest eigenvalue of $\mathbf{V}$ which is given in the next claim.

Theorem 4.9

The largest eigenvalue of $\mathbf{V}$ *is*

$$\widetilde{\lambda}_{max} = e^{(\ell_1 + \cdots + \ell_{2n-1})/2}.$$

Proof. We proceed by steps.

- Let $\mathbf{K}_1, \mathbf{K}_2 \in \mathbf{GL}(2^n, \mathbb{C})$ be two matrices which put

$$\frac{1 + \widetilde{\mathbf{U}}}{2}\,\widetilde{\mathbf{S}}_+ = \mathrm{diag}(\mathbf{A}_+, \mathbf{0}), \qquad \frac{1 - \widetilde{\mathbf{U}}}{2}\,\widetilde{\mathbf{S}}_- = \mathrm{diag}(\mathbf{0}, \mathbf{B}_-),$$

in diagonal form:

$$\mathbf{V}_+ := \mathbf{K}_1\left(\frac{1 + \widetilde{\mathbf{U}}}{2}\,\widetilde{\mathbf{S}}_+\right)\mathbf{K}_1^{-1}, \qquad \mathbf{V}_- := \mathbf{K}_2\left(\frac{1 - \widetilde{\mathbf{U}}}{2}\,\widetilde{\mathbf{S}}_-\right)\mathbf{K}_2^{-1}, \qquad (4.76)$$

- where $\mathbf{V}_\pm$ are diagonal matrices with half of the eigenvalues of $\mathbf{S}_+$ and half that of $\mathbf{S}_-$. Such eigenvalues are exactly the diagonal entries of $\mathbf{V}_\pm$.

- From now on we consider only $\mathbf{V}_-$. Similar considerations hold for $\mathbf{V}_+$. From (4.76) we obtain

$$
\begin{aligned}
\mathbf{V}_- &= \frac{1}{2}\left(\mathbf{K}_2\,\widetilde{\mathbf{S}}_-\,\mathbf{K}_2^{-1} - \mathbf{K}_2\,\widetilde{\mathbf{U}}\,\mathbf{K}_2^{-1}\mathbf{K}_2\,\widetilde{\mathbf{S}}_-\mathbf{K}_2^{-1}\right) \\
&= \frac{1 - \mathbf{K}_2\,\widetilde{\mathbf{U}}\,\mathbf{K}_2^{-1}}{2}\,\mathbf{K}_2\,\widetilde{\mathbf{S}}_-\,\mathbf{K}_2^{-1}.
\end{aligned}
$$

- Recall that $\widetilde{\mathbf{U}} = \mathrm{diag}(1,-1)$ and note that $\mathbf{K}_2\,\widetilde{\mathbf{S}}_-\,\mathbf{K}_2^{-1}$ has the same eigenvalues of $\mathbf{S}_-$, given by (4.64). We now impose that $\mathbf{K}_2\,\widetilde{\mathbf{U}}\,\mathbf{K}_2^{-1}$ remains diagonal (hence the effect of $\mathbf{K}_2$ is just to permute the eigenvalues of ± 1 of $\widetilde{\mathbf{U}}$ along the diagonal) so that the effect of $(1 - \mathbf{K}_2\,\widetilde{\mathbf{U}}\,\mathbf{K}_2^{-1})/2$ on $\mathbf{K}_2\,\widetilde{\mathbf{S}}_-\,\mathbf{K}_2^{-1}$, which must be diagonal with eigenvalues (4.64) on the diagonal, is to eliminate half of these eigenvalues keeping only those which fall into the upper $2^{n-1} \times 2^{n-1}$ square. A direct check shows that this happens if $\mathbf{K}_2\,\widetilde{\mathbf{U}}\,\mathbf{K}_2^{-1} = -\widetilde{\mathbf{U}}$ and if the diagonal matrix $\mathbf{K}_2\,\widetilde{\mathbf{S}}_-\,\mathbf{K}_2^{-1}$ is such that its 2^n diagonal entries are $e^{(\pm\ell_0\pm\ell_2\pm\ell_4\pm\cdots\pm\ell_{2n-2})/2}$, with an even number of $-$ signs appearing in the exponents.

- The same reasoning can be done for $\mathbf{V}_+$. To summarize: half of the eigenvalues of $\mathbf{V}$ are of the form $e^{(\pm\ell_0\pm\ell_2\pm\ell_4\pm\cdots\pm\ell_{2n-2})/2}$, the other half of the form $e^{(\pm\ell_1\pm\ell_3\pm\ell_5\pm\cdots\pm\ell_{2n-1})/2}$. In each eigenvalue an even number of minus signs appears in the exponent.

- We know that for $k = 0,1,\ldots,2n-1$ one has $\ell_k = \ell_{2n-k}$ and $0 < \ell_0 < \ell_1 < \cdots < \ell_n$. We conclude that the largest eigenvalue of $\mathbf{V}$ is

$$
\widetilde{\lambda}_{\max} = e^{(\ell_1 + \cdots + \ell_{2n-1})/2}.
$$

The Theorem is proved. ∎

4.5.4 *The case $H = 0$. Partition function in the thermodynamic limit*

▶ We can now claim that all eigenvalues of $\mathbf{V}$ are positive, the largest one being

$$
\widetilde{\lambda}_{\max} = e^{(\ell_1 + \cdots + \ell_{2n-1})/2}.
$$

Our original problem about the existence and the computation of the limit (4.58),

$$
\lim_{D\to+\infty} \frac{1}{D}\log \mathcal{Z}(\beta,\varepsilon,0) = \frac{1}{2}\log\left(2\sinh(2\varepsilon)\right) + \lim_{n\to+\infty} \frac{1}{n}\log\widetilde{\lambda}_{\max},
$$

is a well-posed problem. It remains to evaluate the eigenvalue $\widetilde{\lambda}_{\max}$. We give the following statement.

Theorem 4.10

There holds

$$\lim_{n \to +\infty} \frac{1}{n} \log \tilde{\lambda}_{\max} = \frac{1}{2} \log \left(\frac{4}{m} \right) + \frac{1}{2\pi} \int_0^\pi \log \left(\frac{1 + \sqrt{1 - m^2 \sin^2 \eta}}{2} \right) d\eta,$$

where

$$m := \frac{2}{\cosh(2\,\varepsilon)\coth(2\,\varepsilon)} = \frac{2\sinh(2\,\beta\,J)}{\cosh^2(2\,\beta\,J)}. \tag{4.77}$$

Proof. We proceed by steps.

- We need to evaluate the quantity

$$L := \lim_{n \to +\infty} \frac{1}{2n} \sum_{k=1}^n \ell_{2k-1}, \tag{4.78}$$

where ℓ_k is the positive solution of (see (4.62))

$$\cosh \ell_k = \cosh(2\,\varepsilon)\cosh(2\,\theta) - \cos\left(\frac{\pi k}{n} \right) \sinh(2\,\varepsilon)\sinh(2\,\theta).$$

- As $n \to +\infty$ we can put (4.78) in integral form. To do so we define a function $x \mapsto \ell(x)$, where $x := \pi(2k - 1)/n$, so that, if $n \to +\infty$, x is a continuous variable and

$$L = \frac{1}{4\pi} \int_0^{2\pi} \ell(x)\, dx = \frac{1}{2\pi} \int_0^\pi \ell(x)\, dx, \tag{4.79}$$

where $\ell(x)$ is the positive solution of

$$\cosh \ell(x) = \cosh(2\,\varepsilon)\cosh(2\,\theta) - \cos x \sinh(2\,\varepsilon)\sinh(2\,\theta). \tag{4.80}$$

The last step in (4.79) is justified by noticing that $\ell(x) = \ell(2\pi - x)$.

- Remember that $\tanh \theta := e^{-2\,\varepsilon}$. This implies

$$\sinh(2\,\theta) = \frac{1}{\sinh(2\,\varepsilon)}, \qquad \cosh(2\,\theta) = \cotanh(2\,\varepsilon).$$

The above formulas allow us to write (4.80) as

$$\cosh \ell(x) = \frac{2}{m} - \cos x,$$

where m is defined as in (4.77).

- We recall the following identity:

$$z = \frac{1}{\pi} \int_0^\pi \log\left(2\cosh z - 2\cos y\right) dy, \qquad z > 0. \tag{4.81}$$

This allow us to write an integral representation of the function ℓ:

$$
\begin{aligned}
\ell(x) &= \frac{1}{\pi} \int_0^\pi \log\left(2\cosh \ell(x) - 2\cos y\right) dy \\
&= \frac{1}{\pi} \int_0^\pi \log\left(\frac{1}{m} - 2\cos x - 2\cos y\right) dy. \tag{4.82}
\end{aligned}
$$

- Substituting (4.82) into (4.79) we get

$$L = \frac{1}{2\pi^2} \int_0^\pi dx \int_0^\pi dy \, \log\left(\frac{1}{m} - 2\left(\cos x + \cos y\right)\right). \tag{4.83}$$

- Symmetry arguments suggest the following change of coordinates:

$$(x, y) \mapsto (\xi, \eta) := \left(\frac{1}{2}(x+y), x-y\right) \in [0, \pi] \times [0, \pi].$$

Then we can write (4.83) in the following form:

$$
\begin{aligned}
L &= \frac{1}{2\pi^2} \int_0^\pi d\xi \int_0^\pi d\eta \, \log\left(\frac{1}{m} - 4\cos\xi \cos\left(\frac{\eta}{2}\right)\right) \\
&= \frac{1}{\pi^2} \int_0^\pi d\xi \int_0^{\pi/2} d\eta \, \log\left(\frac{1}{m} - 4\cos\xi \cos\eta\right) \\
&= \frac{1}{\pi^2} \int_0^\pi d\xi \int_0^{\pi/2} d\eta \, \log\left(2\cos\eta\right) \\
&\quad + \frac{1}{\pi^2} \int_0^\pi d\xi \int_0^{\pi/2} d\eta \, \log\left(\frac{2}{m\cos\eta} - 2\cos\xi\right) \\
&= \frac{1}{\pi} \int_0^{\pi/2} \log\left(2\cos\eta\right) d\eta + \frac{1}{\pi} \int_0^{\pi/2} \cosh^{-1}\left(\frac{1}{m\cos\eta}\right) d\eta,
\end{aligned}
$$

where we used the identity (4.81) to transform the second term in the last step.

- Since

$$\cosh^{-1} z = \log\left(z + \sqrt{z^2 - 1}\right),$$

we obtain

$$L = \frac{1}{\pi} \int_0^{\pi/2} \log\left(\frac{2}{m}\left(1 + \sqrt{1 - m^2\cos^2\eta}\right)\right) d\eta.$$

- A further change of variable and a straightforward manipulation gives the desired formula.

The Theorem is proved. ∎

4.5.5   *The case $H = 0$. Thermodynamics*

▶ Theorem 4.10 gives the final answer we were looking for. We can claim that the TL of the logarithm of the partition function of the two-dimensional Ising model without magnetic field is:

$$\lim_{D \to +\infty} \frac{1}{D} \log \mathcal{Z}(\beta, \varepsilon, 0) = \log (2 \cosh(2\varepsilon))$$

$$+ \frac{1}{2\pi} \int_0^\pi \log \left(\frac{1 + \sqrt{1 - m^2 \sin^2 \eta}}{2} \right) d\eta.$$

▶ Such result allows us to determine the thermodynamics of the two-dimensional Ising model without magnetic field.

Theorem 4.11

> *Consider a two-dimensional Ising model with configurational energy (4.33) with $H = 0$ and in the TL. Then:*
>
> 1. *The free energy per spin is*
>
> $$F = -\frac{1}{\beta} \left(\log (2 \cosh(2\beta J)) + \frac{1}{2\pi} \int_0^\pi \log \left(\frac{1 + \sqrt{1 - m^2 \sin^2 \eta}}{2} \right) d\eta \right),$$
> $$(4.84)$$
>
> *where m is defined in (4.77).*
>
> 2. *The (average) energy per spin is*
>
> $$E = -J \coth(2\beta J) \left(1 + \frac{2 m'}{\pi} \mathcal{I}_1(m) \right),\qquad (4.85)$$
>
> *where $\mathcal{I}_1(m)$ is the complete elliptic integral of the first kind:*
>
> $$\mathcal{I}_1(m) := \int_0^{\pi/2} \frac{d\eta}{\sqrt{1 - m^2 \sin^2 \eta}},\qquad (4.86)$$
>
> *and*
> $$(m')^2 := 1 - m^2 = \left(2 \tanh^2(2\beta J) - 1 \right)^2.$$
>
> 3. *The heat capacity per spin is*
>
> $$C = \frac{2\kappa}{\pi} (\beta J \coth(2\beta J))^2$$
> $$\times \left(2 \left(\mathcal{I}_1(m) - \mathcal{I}_2(m) \right) - (1 - m') \left(\frac{\pi}{2} + m' \mathcal{I}_1(m) \right) \right), \quad (4.87)$$

where $\mathcal{I}_2(m)$ is the complete elliptic integral of the second kind:

$$\mathcal{I}_2(m) := \int_0^{\pi/2} \sqrt{1 - m^2 \sin^2 \eta} \, d\eta. \tag{4.88}$$

Proof. All formulas can be proved by applying definitions of thermodynamic quantities.

1. The free energy per spin is by definition

$$F := -\frac{1}{\beta} \lim_{D \to +\infty} \frac{1}{D} \log \mathcal{Z}(\beta, J, 0),$$

which immediately gives the desired formula if we use our expression of the partition function under the TL.

2. By using definition (4.15) we have

$$E := -\frac{\partial}{\partial \beta} \lim_{D \to +\infty} \frac{1}{D} \log \mathcal{Z}(\beta, J, 0).$$

A direct computation gives

$$E = -2J \tanh(2\beta J) + \frac{m}{2\pi} \frac{\partial m}{\partial \beta} \int_0^\pi \frac{\sin^2 \eta}{\sqrt{1 - m^2 \sin^2 \eta} \left(1 + \sqrt{1 - m^2 \sin^2 \eta}\right)} \, d\eta.$$

Now we have

$$\frac{\partial m}{\partial \beta} = -2J m \coth(2\beta J) \left(2 \tanh^2(2\beta J) - 1\right) = -2J m m' \coth(2\beta J),$$

with $(m')^2 := 1 - m^2 = \left(2 \tanh^2(2\beta J) - 1\right)^2$, and

$$\int_0^\pi \frac{\sin^2 \eta}{\sqrt{1 - m^2 \sin^2 \eta} \left(1 + \sqrt{1 - m^2 \sin^2 \eta}\right)} \, d\eta = \frac{1}{m^2} \left(-\pi + 2\mathcal{I}_1(m)\right),$$

where $\mathcal{I}_1(m)$ is the complete elliptic integral of the first kind, see (4.86). Then we have:

$$
\begin{aligned}
E &= -2J \tanh(2\beta J) + J m' \coth(2\beta J) - \frac{2J m'}{\pi} \coth(2\beta J) \mathcal{I}_1(m) \\
&= -J \coth(2\beta J) \left(1 + \frac{2m'}{\pi} \mathcal{I}_1(m)\right),
\end{aligned}
$$

which gives the desired formula.

3. By using definition (4.16) we have

$$C := \frac{\partial E}{\partial T} = -\kappa \beta^2 \frac{\partial E}{\partial \beta}.$$

Let $\mathcal{I}_2(m)$ be the complete elliptic integral of the second kind, see (4.88). Taking into account that

$$m \frac{\partial}{\partial m} \mathcal{I}_1(m) = \frac{\mathcal{I}_2(m)}{1 - m^2} - \mathcal{I}_1(m),$$

a straightforward computation gives the desired expression.

The Theorem is proved.   ■

▶ We have now all ingredients to prove the existence of a phase transition point and to locate it. Such point is exactly given by the singularity point of the thermodynamic functions given in Theorem 4.11.

Theorem 4.12

Consider a two-dimensional Ising model with configurational energy (4.33) with $H = 0$ and in the TL. Then:

1. *The free energy per spin (4.84) has a singularity at the critical temperature $T = T_c > 0$, where T_c is uniquely determined by*

$$T_c = \frac{2J}{\kappa \log(1 + \sqrt{2})}. \tag{4.89}$$

At $T = T_c$ the free energy per spin fails to be an analytic function of T.

2. *The (average) energy per spin (4.85) is continuous at $T = T_c$:*

$$E(T_c) = -\sqrt{2} J.$$

3. *The heat capacity per spin (4.87) has a logarithmic divergence at $T = T_c$,*

$$C(T_c) \approx R_1 \left(\log \left| \frac{T}{T_c} - 1 \right| + R_2 \right),$$

where R_1 and R_2 are two constants given by

$$R_1 := -\frac{2\kappa \log^2(1 + \sqrt{2})}{\pi},$$

$$R_2 := 1 + \frac{\pi}{4} + \log \left(\frac{\sqrt{2}}{4} \log(1 + \sqrt{2}) \right).$$

Proof. We prove only the last two claims. We first note that condition (4.89) is equivalent to

$$\sinh(2\beta_c J) = 1, \qquad \beta_c := \frac{1}{\kappa T_c},$$

which implies $m = 1$.

2. We know that the (average) energy per spin is

$$E = -J \coth(2\beta J) \left(1 + \frac{2m'}{\pi} \mathcal{I}_1(m) \right), \tag{4.90}$$

We have to approximate E when $T \approx T_c$. To do this we recall that

$$\lim_{m \to 1} \mathcal{I}_1(m) = +\infty, \qquad \lim_{m \to 1} \mathcal{I}_2(m) = 1. \tag{4.91}$$

Then rewrite $\mathcal{I}_1(m))$ as

$$
\begin{aligned}
\mathcal{I}_1(m) &= \int_0^{\pi/2} \frac{d\eta}{\sqrt{\cos^2 \eta + (m')^2 \sin^2 \eta}} \\
&= \int_{\pi/2-\rho}^{\pi/2} \frac{d\eta}{\sqrt{\cos^2 \eta + (m')^2 \sin^2 \eta}} \\
&\quad + \int_0^{\pi/2-\rho} \frac{d\eta}{\sqrt{\cos^2 \eta + (m')^2 \sin^2 \eta}},
\end{aligned} \tag{4.92}
$$

where we assume that ρ satisfies $\rho/|m'| \ll 1$ and $1/\rho \gg 1$. In the first integral of (4.92) define $t := \pi/2 - \eta$. Then, since $\rho \ll 1$,

$$
\begin{aligned}
\int_{\pi/2-\rho}^{\pi/2} \frac{d\eta}{\sqrt{\cos^2 \eta + (m')^2 \sin^2 \eta}} &\approx \int_0^{\rho} \frac{dt}{\sqrt{(m')^2 + m^2 t^2}} \\
&= \frac{1}{m} \log \left(\frac{m\rho + \sqrt{(m')^2 + m^2\rho^2}}{|m'|} \right) \\
&\approx \log \left(\frac{2\rho}{|m'|} \right),
\end{aligned} \tag{4.93}
$$

because $\rho/|m'| \ll 1$. In the second integral of (4.92), $m' \sin \eta$ may be neglected in comparison with $\cos \eta$ because $\eta < \pi/2 - \rho$, so that

$$
\begin{aligned}
\int_0^{\pi/2-\rho} \frac{d\eta}{\sqrt{\cos^2 \eta + (m')^2 \sin^2 \eta}} &\approx \int_0^{\pi/2-\rho} \frac{d\eta}{\cos \eta} \\
&= \log \left(\frac{1}{\sin(\pi/2 - \rho)} + \tan \left(\frac{\pi}{2} - \rho \right) \right) \\
&\approx \log \left(\frac{2}{\rho} \right).
\end{aligned} \tag{4.94}
$$

Combining (4.93) and (4.94) we find that, as $m \approx 1$,

$$\mathcal{I}_1(m) \approx \log\left(\frac{4}{|m'|}\right).$$

Somewhat more precisely, it can be shown that, as $m \approx 1$,

$$\mathcal{I}_1(m) = \log\left(\frac{4}{|m'|}\right) + O\left((m')^2 \log|m'|\right). \tag{4.95}$$

Approximations (4.91) and (4.95) allow us to study the behavior of E if $T \approx T_c$. As $T \approx T_c$, we have:

$$m \approx 1 - 4\,\beta_c^2\, J^2 \left(\frac{T}{T_c} - 1\right)^2,$$

from which one can compute also an approximation for m'. Therefore, from (4.90) we see that E is a continuous function of T even at T_c, where its values is

$$E(T_c) = -\frac{J}{\tanh(2\,\beta_c\, J)} = -\sqrt{2}\, J.$$

3. From the expression of C in (4.87) and from our previous approximations, we have, as $T \approx T_c$,

$$\frac{C}{\kappa} \approx \frac{8\,\beta_c^2\, J^2}{\pi}\left(\log\left(\frac{4}{|m'|}\right) - 1 - \frac{\pi}{4}\right) \approx R_1\left(\log\left|\frac{T}{T_c} - 1\right| + R_2\right), \tag{4.96}$$

where R_1 and R_2 are two constants given by

$$R_1 := -\frac{2\,\kappa\,\log^2(1 + \sqrt{2})}{\pi},$$

$$R_2 := 1 + \frac{\pi}{4} + \log\left(\frac{\sqrt{2}}{4}\log(1 + \sqrt{2})\right).$$

From (4.96) we see that C has a logarithmic divergence at $T = T_c$. Note that R_1 and R_2 are the same for T above and below T_c.

The last two claims are proved.   ■

▶ Remarks:

- In figure 4.7 we see the heat capacity per spin and the (average) energy per spin plotted against the temperature.

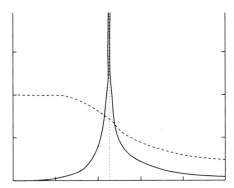

Fig. 4.7. Heat capacity per spin (continuous line) and (average) energy per spin (broken line) plotted against the temperature ([LaBe])

- The behaviors of E and C are qualitatively the same also in the case $J_1 \neq J_2$. In particular, E remains a continuous function of T even at T_c and C has a logarithmic divergence at T_c. The critical temperature is now uniquely determined by the condition

$$\sinh(2\,\beta_c\,J_1)\sinh(2\,\beta_c\,J_2) = 1. \tag{4.97}$$

▶ Note that to examine the *spontaneous magnetization* phenomenon at $T = T_c$ we should repeat the computation of the partition function considering $H \neq 0$. Indeed we should compute the magnetization per spin

$$M := -\frac{\partial F}{\partial H},$$

and notice that in the limit $H \to 0$ such quantity does not vanish. The result is:

$$\lim_{H \to 0} M = \begin{cases} 0 & \text{if } T > T_c, \\ \left(1 - \dfrac{1}{\sinh^4(2\,\beta\,J)}\right)^{1/8} & \text{if } T < T_c. \end{cases}$$

Such result confirms that the two-dimensional Ising models exhibits spontaneous magnetization for $T < T_c$.

4.6   Exercises

Ch4.E1 Determine explicitly the canonical partition function for the following one-dimensional spin chains (at temperature T):

(a) n spins $\omega_i = \pm 1/2$, $i = 1, \ldots, n$, with configurational energy

$$\mathcal{E}(\{\omega\}) := -H \sum_{i=1}^{n} \omega_i, \qquad H > 0.$$

(b) n spins $\omega_i = 0, \pm 1$, $i = 1, \ldots, n$, with configurational energy

$$\mathcal{E}(\{\omega\}) := -H \sum_{i=1}^{n} \omega_i, \qquad H > 0.$$

(c) Three spins $\omega_i = \pm 1/2$, $i = 1, 2, 3$, with configurational energy

$$\mathcal{E}(\{\omega\}) := -J(\omega_1 \omega_2 + \omega_2 \omega_3 + \omega_3 \omega_1) - H(\omega_1 + \omega_2 + \omega_3), \qquad J, H > 0.$$

(d) $n + 1$ spins $\omega_i = \pm 1$, $i = 0, \ldots, n$, with configurational energy

$$\mathcal{E}(\{\omega\}) := -J \sum_{i=1}^{n} \omega_i \omega_0 - H \sum_{i=0}^{n} \omega_i, \qquad J, H > 0.$$

(e) Two different spins $\omega_1 = 0, \pm 1$ and $\omega_2 = \pm 1$, with configurational energy

$$\mathcal{E}(\{\omega\}) := -J \omega_1 \omega_2 - H(\omega_1 + \omega_2), \qquad J, H > 0.$$

Ch4.E2 Consider an open one-dimensional spin chain (at temperature T) with n spins and configurational energy

$$\mathcal{E}(\{\omega\}) := -H \sum_{i=1}^{n} \omega_i, \qquad H > 0,$$

with $\omega_i = -m, -m+1, \ldots, m-1, m$, where $m = (2\ell + 1)/2$, $\ell \in \mathbb{N}$ (i.e., m is a half-odd integer). Determine the partition function as a function of m.

Ch4.E3 Consider a free one-dimensional spin chain with n spins $\omega_i = \pm 1$ and configurational energy

$$\mathcal{E}(\{\omega\}) := -a \sum_{i=1}^{n-1} \omega_i \omega_{i+1} - b \sum_{i=1}^{n-2} \omega_i \omega_{i+2}, \qquad a, b > 0.$$

The system is at temperature T.

(a) Define the new variables

$$t_0 := \omega_1, \qquad t_i := \omega_i \omega_{i+1}, \qquad i = 1, \ldots, n-1.$$

Show that this transformation can be uniquely inverted by finding the inverse transformation $\omega_i = \omega_i(t_0, \ldots, t_{n-1})$.

(b) Prove that the partition function, written in terms of the variables t_i, is proportional to the partition function of a free one-dimensional Ising system with $n-1$ spins with nearest neighbor interactions and magnetic field.

〜〜〜〜〜〜〜〜〜〜〜〜〜〜〜〜〜〜〜〜〜〜〜

Ch4.E4 Consider a one-dimensional spin chain on a closed ring of n sites. Each site supports a spin $\omega_i = \pm 1$, $\omega_{n+1} \equiv \omega_1$. The configurational energy is defined by

$$\mathcal{E}(\{\omega\}) := -J \sum_{i=1}^{n} \left(\frac{1 + \omega_{i-1}\,\omega_i}{2} \right) \left(\frac{1 + \omega_i\,\omega_{i+1}}{2} \right), \qquad J > 0.$$

The system is at temperature T.

(a) Compute explicitly the quantity

$$P(\omega, \omega') := \frac{1 + \omega\,\omega'}{2},$$

where ω, ω' are two spin variables.

(b) Show that the configurational energy is proportional to the number of consecutive triples of sites $(i-1, i, i+1)$ for which $\omega_{i-1} = \omega_i = \omega_{i+1}$.

(c) Determine the partition function for a ring of $n = 4$ sites.

〜〜〜〜〜〜〜〜〜〜〜〜〜〜〜〜〜〜〜〜〜〜〜

Ch4.E5 Consider a closed one-dimensional spin chain (at temperature T) with n spins and configurational energy

$$\mathcal{E}(\{\omega\}) := n\,Q^2 - \sum_{i=1}^{n} \left(A + (-1)^i B\,Q \right) \omega_i\,\omega_{i+1},$$

with $\omega_i = \pm 1$, $\omega_{n+1} \equiv \omega_1$, $A, B, Q > 0$ parameters of the model.

(a) Define the 2×2 matrices

$$T_\pm(\omega_i, \omega_{i+1}) := \exp(\beta\, C_\pm\, \omega_i\, \omega_{i+1}), \qquad \beta := (\kappa\, T)^{-1},$$

with $C_\pm := A \pm B\,Q$. Prove that the partition function can be written as

$$\mathcal{Z} = R \sum_{\omega_1 = \pm 1} \cdots \sum_{\omega_n = \pm 1} \prod_{i=1}^{n/2} T_-(\omega_{2i-1}, \omega_{2i})\,T_+(\omega_{2i}, \omega_{2i+1}),$$

where R is an overall factor to be determined.

(b) Find the matrix τ, defined in terms of the matrices $T_\pm$, such that

$$\mathcal{Z} = R \operatorname{Trace}\left(\tau^{n/2} \right).$$

(c) Find the explicit expression of $\mathcal{Z}$ in terms of the eigenvalues of τ.

(d) Compute $\mathcal{Z}$ in the thermodynamic limit.

〜〜〜〜〜〜〜〜〜〜〜〜〜〜〜〜〜〜〜〜〜〜〜

Ch4.E6 Consider a closed n-site model in which there are three possible states per site, which we can denote by A, B, V, where the states A and B are identical. The energies of the links $A - A$, $A - B$, and $B - B$ are all identical and equal to J. The state V represents a vacancy, and any link containing a vacancy, i.e. $A - V, B - V, V - V$ has energy 0.

(a) Suppose we write $\omega = +1$ for A, $\omega = -1$ for B and $\omega = 0$ for V. Find a function $f(\omega_i, \omega_j)$ such that

$$\mathcal{E}(\{\omega\}) := \sum_{i,j} f(\omega_i, \omega_j) \qquad (4.98)$$

is the configurational energy of the model. Here the sum is over nearest neighbors on the lattice.

(b) Consider a triangle and put a spin at each vertex. The configurational energy is given by (4.98) Find the average total energy at temperature T.

Ch4.E7 Consider a closed n-site model in which there are three possible states per site, which we can denote by A, B, V, where the states A and B are identical. The energies of the links $A - A$, $A - B$, and $B - B$ are all identical and equal to J. The state V represents a vacancy, and any link containing a vacancy, i.e. $A - V, B - V, V - V$ has energy 0.

We write $\omega = +1$ for A, $\omega = -1$ for B and $\omega = 0$ for V. The configurational energy is:

$$\mathcal{E}(\{\omega\}) := J \sum_{i,j} \omega_i^2 \, \omega_j^2,$$

where the sum is over nearest neighbors on the lattice. The system is at temperature T.

(a) Find the transfer matrix of the system.

(b) Find the partition function using the transfer matrix method.

(c) Find the free energy in the thermodynamic limit.

Ch4.E8 Consider a two-dimensional Ising-type model with n sites with two kinds of site, say A and B, as shown in the figure.

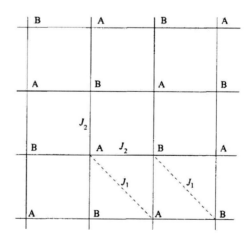

Spins $\omega_i(A), \omega_i(B) \in \{\pm 1\}, i = 1, \ldots, n$, interact in the following way:

- There is an interaction $-J_1\,\omega_i(A)\,\omega_j(A)$, with $J_1 > 0$, between nearest neighbors i and j of A-type.
- There is an interaction $-J_1\,\omega_i(B)\,\omega_j(B)$, with $J_1 > 0$, between nearest neighbors i and j of B-type.
- There is an interaction $+J_2\,\omega_i(A)\,\omega_j(B)$, with $0 < J_2 < J_1$, between nearest neighbors i of A-type and j of B-type.

(a) Determine the configurational energy of the model assuming that there is also an external magnetic field which acts on all the spins in the same way, say $H_A = H_B =: H$.

(b) Determine the partition function assuming that $J_1 = J_2 = 0$.

(c) Let $J_1 = J_2 = 0$ and assume that the magnetic field acts on spins A with a coefficient H_A and on spins B with $H_B \neq H_A$. Determine the partition function.

〜〜〜〜〜〜〜〜〜〜〜〜

Ch4.E9 The two-dimensional Ising model is quite a convincing model for binary alloys. Consider a square lattice of atoms, which can be either of type A or B.

B	B	B	B
B	B	A	A
B	A	A	B
A	B	A	B

↓	↓	↓	↓
↓	↓	↑	↑
↓	↑	↑	↓
↑	↓	↑	↓

We make the following identification: $A \equiv +1$ and $B \equiv -1$. Let the number of the two kinds of atoms be N_A and N_B, with $N_A + N_B = N$, let the interaction energies (bond strengths) between two neighboring atoms be J_{AA}, J_{BB} and J_{AB}. Let the total number of nearest-neighbor bonds of the three possible types be N_{AA}, N_{BB} and N_{AB}. Then the configurational energy for the binary alloy is

$$\mathcal{E}_{\text{binary}} := -J_{AA}\,N_{AA} - J_{BB}\,N_{BB} - J_{AB}\,N_{AB}.$$

Prove that $\mathcal{E}_{\text{binary}}$ can be identified with the configurational energy of a two-dimensional Ising model with N sites (with nearest neighbor interactions):

$$\mathcal{E}_{\text{Ising}} := -\alpha\,N - J\sum_{i,j}\omega_i\,\omega_j - H\sum_i \omega_i,$$

where α, E, H are parameters, depending on J_{AA}, J_{BB} and J_{AB}, to be determined.

Index